Tissue Engineering

Jeong-Yeol Yoon

Tissue Engineering

A Primer with Laboratory Demonstrations

 Springer

Jeong-Yeol Yoon
Department of Biomedical Engineering
The University of Arizona
Tucson, AZ, USA

ISBN 978-3-030-83695-5   ISBN 978-3-030-83696-2   (eBook)
https://doi.org/10.1007/978-3-030-83696-2

This Springer imprint is published by the registered company Springer Nature Switzerland AG
The registered company address is: Gewerbestrasse 11, 6330 Cham, Switzerland

Preface

I have been teaching a semester-long upper-division undergraduate college class on tissue engineering. There are several good textbooks, albeit not many, on tissue engineering. (This is understandable since tissue engineering is still a new discipline.) However, many textbooks are exceptionally long, close to a thousand pages or even two thousand pages. They are written by many authors, in a manner of an *edited book*, potentially lacking the following: consistency, explanations of basic terms, example questions, and engineering design/calculation.

In addition, I have been a firm believer that the core bioengineering and related-discipline classes are to be delivered together with the hands-on laboratory exercises. I have already demonstrated this concept in my other textbook, *Introduction to Biosensors*, whose second edition was also published by Springer in 2016.

With this book, I offer a single-authored primer textbook on tissue engineering, not in the manner of a lengthy edited book. It covers the fundamental basics of tissue engineering in a concise manner, accompanied by a series of laboratory exercises. While we can certainly use this book for a class with laboratory exercises, we can also use it without laboratories. A list of questions and discussion topics are added to each laboratory exercise. The instructor can lead such discussion collaboratively using the experimental procedures and results included in the textbook. As the laboratory results are included with actual images and data for all laboratory exercises, scientists and engineers not in colleges can also learn the concepts in a hands-on and visual manner to better understand concepts. It will also lay a foundation to build their experiments towards their research and commercial development.

With this book, I aimed to accomplish the followings:

- Most up-to-date aspects of tissue engineering are covered in a concise manner while providing easy-to-understand basic concepts.
- Step-by-step learning of all necessary concepts is provided.
- A large number of figures are provided. There are 226 figures through 14 chapters in this book.
- Simple, low-cost, and easy-to-implement laboratory exercises are included in all chapters, except for the first and last chapters.

– Photographs of all equipment and results of all laboratories are provided to visualize tissue engineering concepts. These can be used as guides for practical laboratory exercises towards better understanding the concepts or starting points towards further research and development.

I sincerely thank my graduate teaching assistants at the University of Arizona. They have helped me create and implement the laboratory exercises in my Cell and Tissue Engineering class in the spring semesters from 2015 to 2021. I specifically appreciate the following graduate teaching assistants: Dr. Katherine Klug (currently at Davids Engineering), Dr. Soohee Cho (currently at Abbott), Dr. Tiffany-Heather Ulep (currently at Roche), and Dr. Kattika Kaarj (presently a faculty member at Mahidol University in Thailand), who had laid foundations in these laboratory exercises. The following graduate teaching assistants have also helped establish the laboratory exercises: Dr. Soo Chung (presently at United States Department of Agriculture – Agricultural Research Service), Mr. Kenneth Schackart (currently at the University of Arizona), and Mr. Ryan Zenhausern (currently at Georgia Institute of Technology). Other graduate teaching assistants, Mr. Christopher Camp and Ms. Carissa Grijalva, both at the University of Arizona, have also contributed substantially to the laboratory exercises. I also thank all students who took my cell and tissue engineering class for providing critical feedback and corrections.

I also thank my former and current department heads, Dr. Urs Utzinger and Dr. Arthur Gmitro in the Biomedical Engineering Department. They have supported my class by providing personnel, equipment, and laboratory space at the University of Arizona. Support and suggestions from the editorial office at Springer, especially Michael McCabe, are greatly appreciated. Finally, I want to sincerely thank my wife, Dr. Sunhi Choi (Mathematics Department at the University of Arizona), for her continuous inspiration and support during the preparation of this textbook.

Tucson, AZ, USA   Jeong-Yeol Yoon

Contents

About the Author

Jeong-Yeol Yoon received his B.S., M.S., and Ph.D. degrees in chemical engineering from Yonsei University, Seoul (South Korea), in 1992, 1994, and 1999, respectively. His dissertation advisor was Dr. Woo-Sik Kim, and his co-advisor was Dr. Jung-Hyun Kim, while he worked primarily on polymer colloids. Dr. Yoon received his second Ph.D. degree in biomedical engineering from the University of California, Los Angeles, in 2004, working on lab-on-a-chip and biomaterials under the guidance of Dr. Robin L. Garrell. He joined faculty at the University of Arizona in August 2004 and currently holds split home appointments in the Department of Biomedical Engineering (primary) and Department of Biosystems Engineering (secondary). Dr. Yoon also holds joint appointments in Chemistry & Biochemistry and BIO5 Institute. He is also associate head for graduate affairs in the Department of Biomedical Engineering, starting from July 2018. He is currently directing Biosensors Lab. Dr. Yoon is a member of IBE, ASABE, SPIE, BMES, and ACS and was councillor-at-large for IBE for the 2010 and 2011 calendar years. He was the president of the Institute of Biological Engineering (IBE) in 2015. Dr. Yoon currently serves as editor-in-chief for the *Journal of Biological Engineering* (the official journal of IBE), associate editor for *Biosensors and Bioelectronics* (Elsevier), and editorial board member for Scientific Reports (Springer Nature) and *Micromachines* (MDPI). Dr. Yoon is the sole author of another Springer book, *Introduction to Biosensors – From Electric Circuits to Immunosensors*, second edition, published in 2016, written similarly to this book.

Chapter 1
Introduction

In this chapter, we will learn several different definitions of tissue engineering as well as its applications.

Inquiry 1. In your own words, define tissue engineering.

Inquiry 2. What can we do with tissue engineering?

1.1 Narrow Definition of Tissue Engineering

When a tissue or an organ starts malfunctioning and drugs cannot resolve such problems, you may wish to "substitute" the damaged tissue or organ with a new one. Tissue or organ can be harvested from a human donor, either deceased or alive, and transplanted to the patient. This procedure is called *organ transplantation*. While it has successfully been practiced in the past, you have probably heard about the extremely long waiting list of organ transplantation, and many patients would die before receiving it. Because of this difficulty, synthetic materials have been used to replace the damaged tissue or organ, made from metals, polymers, or ceramics. They are called *implants*, which have shown limited successes in organs, such as bones. However, such implants cannot metabolize nutrients, cannot produce proteins, etc., and subsequently cannot truly replace the functions of tissues or organs.

Therefore, a new cell-based approach is needed to replicate a living tissue's or organ's structure and function. Tissue or organ can be damaged from traumatic injury, degradation from exercise, cancer, aging, and adult-onset deficiency such as diabetes. As tissue is made from cells and *extracellular matrix* (*ECM*), it is important to replicate both the cells and ECM. (Implants may replace the ECM but not the cells.) Both the cells and ECM must be "engineered" to provide structural integrity and metabolic behavior found in normal tissue. With the narrow definition of tissue engineering, engineered mimic of the ECM is constructed, made from either a natural material (e.g., collagen fibers) or a synthetic material (e.g., polymers). This ECM

© Springer Nature Switzerland AG 2022
J.-Y. Yoon, *Tissue Engineering*, https://doi.org/10.1007/978-3-030-83696-2_1

mimic is called *tissue-engineered scaffold* (*TE scaffold*) or just *scaffold*. Cells are then harvested from a patient or a donor. We can also use *stem cells*. They are cultured and seeded on the scaffold in a laboratory, that is, in vitro (meaning "in a representation," typically representing a dish, flask, beaker, tube, etc.). The antonym to in vitro is in vivo, meaning "in life," typically representing a (human) body. Sometimes the term ex vivo is used as an antonym to in vivo. While in vitro and ex vivo do not necessarily represent the same, these two terms are sometimes used interchangeably. Once the cells proliferate and exhibit normal metabolism in a scaffold, it is then transplanted back to the body to replace the tissue or organ. This final product is called *tissue-engineered transplant* (*TE transplant*). This procedure describes the narrow definition of tissue engineering.

Early successes in tissue-engineered constructs date back to the late 1990s. The most famous example would be the work by Cao et al., published in 1997, where a tissue-engineered human ear was constructed on a nude mouse [Cao et al., 1997; Bear in mind that there were many limitations and issues with this early pioneering work]. Since then, tissue engineering has become widely known and practiced. In the early 2000s, it has been established as a discipline. Tissue engineering is still a new discipline at the time of writing.

Figure 1.1 shows one example of creating a tissue-engineered transplant (TE transplant), that is, narrow definition of tissue engineering. Cells are removed from a healthy host (or a healthy tissue from the same patient). They are cultured in a laboratory, and the best-performing cell line is isolated. If stem cells are used, they should also be differentiated using proper differentiation factors, physical stimuli, and other environmental factors. A scaffold is constructed and the chosen cells are seeded. Once the transplant is complete, it is transplanted to the patient.

Tissue-engineered transplants have numerous benefits over implants. Many implants can be rejected from a body via inflammatory and immune responses, where such issue is minimal with TE transplants. Since the best-performing cell line can be selected and the scaffold design can be optimized, the resulting TE transplants can perform far better than implants, while there is no availability issue common in organ transplants.

1.2   Early Attempt in Scaffold Development: Decellularized Matrix

Since its conception in the early 2000s, scaffold design has been considered quite challenging toward fully mimicking the natural ECM. One early attempt was the use of natural ECM rather than designing and fabricating a whole new scaffold. An organ from a dead body (*cadaver*) was harvested and the cells were removed with surfactant (*decellularization*). This *decellularized matrix* was then used as a scaffold. Ott et al. demonstrated this method using the heart from cadaver (published in 2008). After harvesting, they decellularized with surfactant (Fig. 1.2) and added

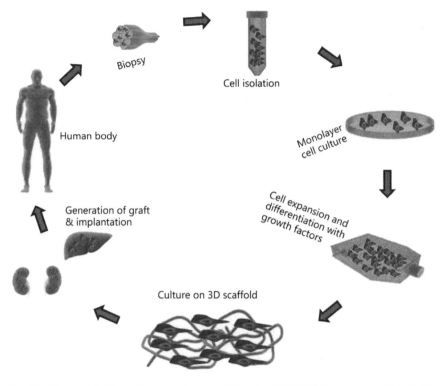

Biopsy

Cell isolation

Human body

Monolayer cell culture

Generation of graft & implantation

Cell expansion and differentiation with growth factors

Culture on 3D scaffold

Fig. 1.1 Narrow definition of tissue engineering. (Ude et al., 2018. (C) Open access article distributed under the terms of the Creative Commons Attribution 4.0 International License)

cardiac cells onto it. Cells were proliferated and maintained for up to 28 days in a bioreactor (in vitro) (Fig. 1.3). The resulting construct generated the pump function equivalent to about 2% of adult heart function. While 2% seemed very low, it was the first demonstration of a thriving TE heart with pump function. It showed the early promise of tissue engineering and has received a lot of media attention.

This method was groundbreaking in many aspects, eliminating many organ transplantation issues, especially tissue healthiness or organ availability. However, it still suffered numerous problems, such as (1) availability of cadavers and (2) lack of engineering control over the scaffold design and structure. Therefore, follow-up works have been focused on the design and construction of "engineered" scaffolds that would meet the requirements of:

- Size (volume) that can "process" the required amounts of nutrient metabolism, protein production, waste removal, etc.
- Mechanical properties that can fit within the available space, resist external bending and shear force, etc.
- Longevity that can last for a necessary duration of time

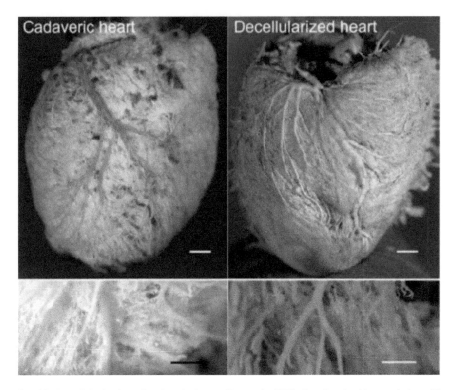

Fig. 1.2 Decellularization of cadaveric heart. (Ott et al., 2008. Reprinted with permission, (C) 2008 Springer Nature)

1.3   Simple TE Transplant Example: Skin

One of the early examples of TE transplants is skin transplant as skin is highly pro-lific and self-renewable, thus relatively easy to replicate with tissue engineering technology. The skin has two layers: the outermost and top layer of the *epidermis* and the immediate inner layer of the *dermis* (just underneath the epidermis) (Fig. 1.4). *Keratinocytes* and *fibroblasts* make up most of the epidermis and dermis, and they proliferate quite well both in vivo and in vitro. Damages of the epidermis can be healed relatively quickly and do not generally require transplants. However, severe burns damage the dermis layer and do not heal very well and usually leave scars, where TE skin transplant can be the right solution. Diabetic patients also have skin ulcer problems that do not heal, called *diabetic ulcers* (a famous example is a *diabetic foot*), where TE skin transplant can also be utilized. TE skin transplant can also be used for cosmetic purposes (plastic surgery) to replace the aged skin with a fresh one.

As skin structure is relatively simple, a simple hydrogel of collagen fibers can be used as a TE skin scaffold. A *gel* is a network of natural or synthetic polymers that are cross-linked, filled with liquid. If the liquid is water, it is specifically called

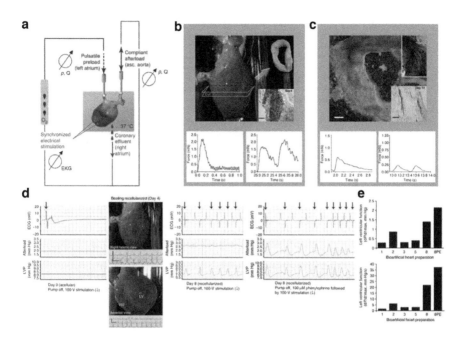

Fig. 1.3 Perfusion bioreactor for constructing a tissue-engineered transplant from the decellularized cadaveric heart. (Ott et al., 2008. Reprinted with permission, (C) 2008 Springer Nature)

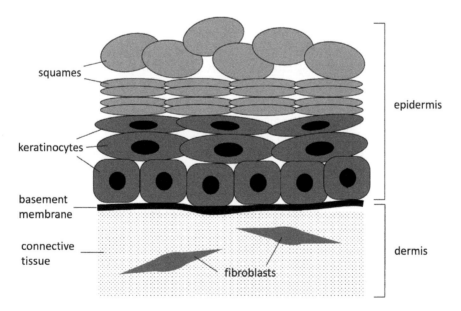

Fig. 1.4 Human skin. The outermost layer is the epidermis (consisting of squames and keratinocytes), followed by the dermis (consisting of fibroblasts and ECM)

Fig. 1.5 Bilayered
tissue-engineered skin
transplant (MyDerm). (Ude
et al., 2018. (C) Open
access article distributed
under the terms of the
Creative Commons
Attribution 4.0
International License)

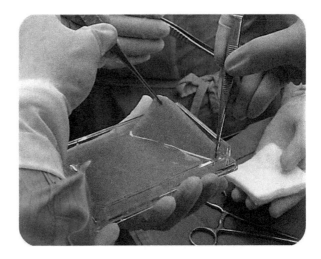

hydrogel. Hydrogels contain as much as >90% water, which is similar to human tissues. Human ECM is made from *collagen fibers*, where the protein *collagen* is polymerized and bundled together to form a highly cross-linked fiber network. Therefore, human ECMs are also hydrogels. Keratinocytes and fibroblasts are seeded on these collagen gels and proliferated in vitro to create a TE skin transplant. Two layers are needed, one with keratinocytes to form the epidermis and the other with fibroblasts to form the dermis (Fig. 1.5). The collagen fibers can be modified with growth factors, drugs, etc., and can also be engineered to exhibit appropriate tissue property (elasticity, tensile strength, etc.).

1.4   Simple TE Transplant Example: Pancreas

Another example is the TE pancreas transplant. The *pancreas* is located just underneath the liver, and one of its primary functions is the release of protein *insulin*, which regulates the glucose level in a body. When there is a problem with insulin, such a disease is called *diabetes*. While drugs and adequate exercise may control the diabetes symptoms, it is often necessary to self-inject insulin daily, sometimes more than once a day. While *continuous glucose monitoring* (*CGM*) system and subsequent automatic insulin injection have become a reality, the ultimate solution would be the use of TE pancreas transplant. *Pancreatic cancer* is another severe problem whose mortality rate remained very high among all cancer types. Again, a TE pancreas transplant may be able to resolve this disease.

Decellularized cadaveric pancreas can be used as a TE scaffold, and the *β-islet cells* (the primary cells in the pancreas) can be seeded and proliferated in vitro. The β-islet cells can be harvested from a healthy donor. Unlike organ transplants, the donor does not need to provide the whole pancreas; a small number of β-islet cells can be extracted and proliferated in a laboratory (in vitro). The resulting TE transplant may be rejected in the long run via immune reaction, and therefore, the immunosuppressant drug should be taken throughout the patient's lifespan. If the donor is very well matched with the patient, such immune rejection could be minimal.

A better alternative is the use of a semipermeable membrane as a TE scaffold. As shown in Fig. 1.6, many different forms of such semipermeable membranes can be used: (A) a cylindrical tube (macrocapsule), (B) a push-pull device with refillable oxygen in it, and (C) a sphere (microcapsule). β-islet cells are loaded inside these devices, where the semipermeable membrane allows the movements of oxygen, nutrients, wastes, and protein products (e.g., insulin), moving freely inside and out. However, these membranes' pores are small enough to keep the β-islet cells inside and not allow the immune cells (white blood cells) to attack the β-islet cells. This procedure is called *immunoisolation*, which has been a popular method in designing a TE scaffold.

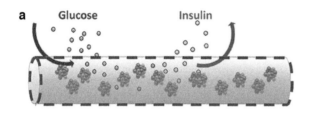

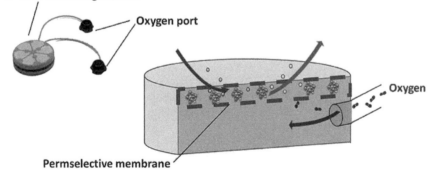

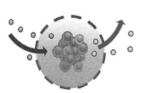

Fig. 1.6 Immunoisolation methods can be used toward TE pancreas transplant. (Hu & de Vos, 2019 (C). Open access article distributed under the terms of the Creative Commons Attribution 4.0 International License)

1.5 Expanded Definition of Tissue Engineering: Organ-on-a-Chip (OOC)

Another emerging application of tissue engineering is *organ-on-a-chip* or simply *organ-on-chip* (*OOC*). Organ-on-a-chip (OOC) is a special case of lab-on-a-chip (LOC). *Lab-on-a-chip* (*LOC*) is a miniaturized laboratory fabricated on a "chip." Chips are made with semiconductor manufacturing technology, for example, photolithography. In recent years, *soft lithography* is more popularly used over photolithography. In both methods, chips are made from silicone or silicone-based polymer, while other polymers can also be used. LOC is essentially a network of wells interconnected with sub-millimeter or micrometer-sized channels. The reagents and specimens are mixed, incubated, reacted, and quantified by various sensors and biosensors in LOCs (Fig. 1.7). LOCs have popularly been used toward point-of-care clinical diagnostics, field-based sensing and biosensing applications, etc.

There has been a growing interest in adapting these LOCs to replace the in vitro cell assays toward improved recapitulation of mammalian tissues. For example, mammalian cells can be added and proliferated within the LOC wells or channels mimicking the actual tissue structures. These are called organ-on-chips (OOCs). Hence, this method utilizes the tissue engineering concept, where the LOC is the TE scaffold. One of the successful demonstrations of OOC is shown in Fig. 1.8, mimicking a human lung (thus *lung-on-a-chip*). A single channel is separated by a membrane, where the lung epithelial cells are seeded and proliferated on it. The bottom channel represents the blood flow, while the top channel represents the air low. A single channel can be utilized to mimic the unit lung behavior, or a collection of channels can be utilized to simulate the overall tissue-level behavior.

Fig. 1.7 Lab-on-a-chip is a network of channels and wells fabricated on silicone or silicone-based polymer. (Picture was taken by Vjsiebens in March 2009 and placed in the public domain. Accessed January 2021 from http://commons. wikimedia.org/wiki/ File:AutoFISH.jpg)

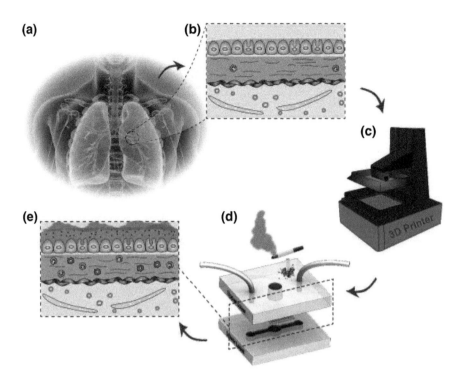

Fig. 1.8 Lung-on-a-chip example. (Shrestha et al., 2019. (C) Open access article distributed under the terms of the Creative Commons Attribution 4.0 International License)

OOCs are typically not for in vivo use. Instead, OOCs provide a laboratory in vitro platform for the following applications:

- Cancer study: *carcinogenesis* (formation of cancer); *metastasis* (spread of cancer), etc.
- Stem cell differentiation study: identification and optimization of physical and biological cues to induce successful differentiation to stem cells
- Biocompatibility test of synthetic materials (e.g., implants)
- Drug efficacy and toxicity tests
- Cosmetics toxicity tests
- Studying the long-term exposure of environmental toxicants to human tissues and organs

The above studies have conventionally been conducted by (1) in vitro cell assays, (2) animal tests, and (3) human clinical trials, in this order. There are huge gaps between (1) and (2) – different responses from individual cells versus tissue or organ – and between (2) and (3) – species difference between animal and human. As human cells are used to create a tissue or organ mimic in OOCs, such gaps would be minimized and eventually eliminated. Besides, as the final device does not need to be transplanted back to humans while there is a huge commercial potential, OOCs have gained substantial popularity in tissue engineering research.

1.6 Overview of This Book

This book is intended for the audience with limited understanding and experience in wet laboratory-based biology. Brief and easy-to-practice laboratory modules are introduced in each chapter. They can certainly be used for an undergraduate-level class (or a graduate-level introductory class) with lecture and laboratory components. However, it will equally be useful for the course with no laboratory components. The laboratory procedures are well-documented, with ample figures, photographs, experimental data, and subsequent analyses, which provides the virtual laboratory experience. It will equally be great for the people self-studying tissue engineering.

Chapter 2 will learn the fundamental basics of cell culture and the basic laboratory skills necessary to conduct it. Bacterial cell culture (the easiest cell culture) laboratory will be introduced, and the use of a microscope will be demonstrated.

Chapter 3 will then advance to cell metabolism – how cells consume nutrients and oxygen and generate wastes. Feeding and passaging – the two most essential skills for mammalian (including human) cell culture – will be demonstrated as laboratory exercises.

In Chapter 4, once cells are successfully cultured, it is crucial to "stain" the different areas of cells and image them. This staining and imaging process is mostly done by fluorescence microscopy. Basic principles of fluorescence, staining, and fluorescence microscopy will be covered, along with the laboratory exercise on fluorescence cell imaging.

Chapter 5 will learn stem cells, which are quite crucial for many tissue engineering applications (although not always necessary). Stem cell culture and confirmation of differentiation will be demonstrated as laboratory exercises.

Chapter 6 will learn various biomaterial surfaces – basically what material is preferred for tissue engineering applications and whether a surface modification is necessary or not. As tissue engineering requires the use of "engineered" scaffold (including organ-on-a-chip), choice and modification of biomaterial surface must be preceded before constructing the scaffold. Several different biomaterial surfaces will be fabricated or prepared in the laboratory exercise. Their surface properties, especially the two most popularly used ones – contact angle and surface roughness – will be evaluated through the laboratory exercise.

Chapter 7 will learn the focal adhesion. It is a primary mechanism for most mammalian (including human) cells to be anchored on the extracellular matrix (ECM) to form a functional tissue, which must be duplicated in tissue-engineered devices. Fluorescence imaging to confirm focal adhesion will be exercised through a laboratory exercise.

Chapter 8 will learn contact guidance. Through focal adhesion, cells can be aligned with the shape and pattern of ECM, which is the definition of contact guidance. Again, contact guidance must be demonstrated in tissue-engineered devices to create a fully functioning tissue-engineered transplant or tissue/organ mimic (OOC). Contact guidance on a simple microgroove will be demonstrated as a laboratory exercise.

In Chapter 9, two popular methods of creating 3D TE scaffold will be explained – electrospinning and 3D printing. Laboratory exercises will be introduced on these two methods.

In Chapter 10, the engineering approach for designing in vitro cell culture and bioreactor will be introduced. Characteristic times will be introduced to design the in vitro culture, including the initial number of cells, time for culture, consideration of Hayflick limit, the volume of a bioreactor, consumption rates of nutrients and oxygen, and the design of TE scaffold based on diffusional lengths, etc. Various types of bioreactors for tissue engineering will also be covered.

In Chapter 11, organ-on-a-chip will be discussed as the first representative application of tissue engineering. Laboratory exercises on organ-on-a-chip fabrication and operation will be introduced.

In Chapter 12, TE skin transplant will be discussed as the second representative application of tissue engineering. A quick laboratory exercise of TE skin transplant will be demonstrated.

In Chapter 13, angiogenesis (formation of a new blood vessel) will be introduced toward vascularization as it is necessary to create a fully functioning TE device. Angiogenesis-on-a-chip will be demonstrated as its laboratory example.

In Chapter 14, advanced topics will be covered, including cartilage tissue engineering, bone marrow transplantation, cardiac patches, immunoisolated pancreas, and kidney tissue engineering.

Figure 1.9 graphically summarizes the overall structure of this book.

Review Questions
1. Compare narrow and broader definitions of tissue engineering (TE).
2. What is extracellular matrix (ECM)?
3. What is TE scaffold?
4. What is hydrogel?
5. What is decellularization? How is it used for tissue engineering applications?
6. Injecting β-islet cells from a well-matched donor to a diabetic patient provides only temporary relief. Why? How can you resolve this issue?
7. What is immunoisolation, and why is it necessary for tissue engineering?
8. How is TE skin transplant made?
9. What are the benefits of organ-on-a-chip (OOC)?

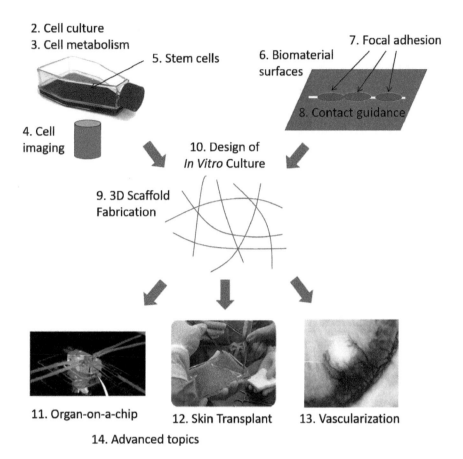

Fig. 1.9 Overview of this book

References

Cao, Y., Vacanti, J. P., Paige, K. T., Upton, J., & Vacanti, C. A. (1997). Transplantation of chondrocytes utilizing a polymer-cell construct to produce tissue-engineered cartilage in the shape of a human ear. *Plastic and Reconstructive Surgery, 100*(2), 297–302. https://doi.org/10.1097/00006534-199708000-00001

Hu, S., & de Vos, P. (2019). Polymeric approaches to reduce tissue responses against devices applied for islet-cell encapsulation. *Frontiers in Bioengineering and Biotechnology., 7*, 134. https://doi.org/10.3389/fbioe.2019.00134

Ott, H. C., Matthiesen, T. S., Goh, S. K., Black, L. D., Kren, S. M., Netoff, T. I., & Taylor, D. A. (2008). Perfusion-decellularized matrix: Using nature's platform to engineer a bioartificial heart. *Nature Medicine, 14*, 213–221. https://doi.org/10.1038/nm1684

Shrestha, J., Ghadiri, M., Shanmugavel, M., Bazaz, S. R., Vasilescu, S., Ding, L., & Warkiani, M. E. (2019). A rapidly prototyped lung-on-a-chip model using 3D-printed molds. *Organs-on-a-Chip, 1*, 100001. https://doi.org/10.1016/j.ooc.2020.100001

Ude, C. C., Miskon, A., Idrus, R. B. H., & Bakar, M. B. A. (2018). Application of stem cells in tissue engineering for defense medicine. *Military Medical Research, 5*, 7. https://doi.org/10.1186/s40779-018-0154-9

Chapter 2
Cell Culture

In the previous chapter, we have learned the overview of tissue engineering principles and their applications. All tissue engineering applications start from growing cells in an appropriate culture, which is the first topic that we should learn. Here are two inquires for you:

Inquiry 1. What is cell culture?

More importantly, how do we "feed" cells and maintain their health? How can we assess their health (typically by monitoring their shape, metabolism, and protein production)?

Inquiry 2. How can we classify cell types?

Different cell types require varied methods of cell culture. Have you heard anchorage-dependent versus anchorage-independent cells? Have you heard primary versus stem cells? How about transformed or immortalized cells?

2.1 What Is Cell Culture?

While there are many different definitions of *cell culture*, we can always find the following three components in its definition:

(1) *In vitro*, meaning "in a representation," that is, in a tube or a culture dish, and outside a living organism (antonym is in vivo)
(2) *Maintain cells in a viable state*, indicating that cells should be alive and actively proliferating (dividing or reproducing)
(3) *Maintain cells in a metabolically active state*

To accomplish all three, we need to feed oxygen, nutrients, and growth factors appropriately. Maintenance of metabolically active state is quite crucial for tissue engineering applications. Cells should produce necessary proteins and respond to

© Springer Nature Switzerland AG 2022
J.-Y. Yoon, *Tissue Engineering*, https://doi.org/10.1007/978-3-030-83696-2_2

environmental factors (e.g., physical and biological cues) to perform essential functions toward tissue engineering applications.

Cells are typically harvested from humans for most tissue engineering applications. For specific tissue engineering applications, we can also use the cells from animals, plants, and bacteria. In many tissue engineering applications, it may be preferable to use a cell population descended from a single cell, referred to as a *cell line*, as they would contain identical genetic information. However, most mammal and human cells would divide only up to a certain number of doublings (known as Hayflick limit, discussed later in Sect. 10.2), and there are limitations in using cell lines. On the other hand, certain cells can divide into an indefinite number of doublings, appropriate as cell lines. These cells are referred to as *immortalized cells*. Cancer cells are probably the most well-known example of immortalized cells (they are spontaneously immortalized cells). One of the earliest human cell lines is *HeLa cells*, descended from the cancer cell from Henrietta Lacks in the 1960s, who died of cancer. Figure 2.1 shows the microscopic image of cultured HeLa cells with *Hoechst staining*, which stains the cell nuclei in blue. (These days, cell nuclei are stained with better dye, DAPI, which will be discussed later in Sect. 4.3.)

Cell cultures are commonly used for various applications. From the definition of cell culture, cells must maintain a metabolically active state in cell culture, producing adequate amounts of proteins and responding to various environmental factors. The protein production feature can be utilized in many different industrial applications. Such proteins are typically used as drugs, including vaccines. A large-scale bioreactor can also be used instead of a simple culture dish to produce many such drugs. Such specific application is often referred to as cellular engineering, which shares many standard features with tissue engineering.

Protein products from cell culture include (1) enzymes, (2) hormones, (3) viral vaccines, (4) antibodies, and (5) cytokines. *Enzymes* are proteins that function as biological catalysts, that is, facilitating and speeding up chemical reactions. Many of them can be utilized as drugs to facilitate or speed up specific chemical reactions.

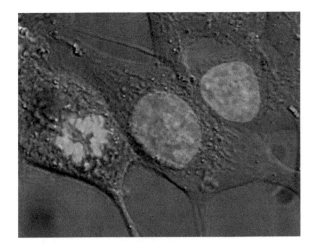

Fig. 2.1 HeLa cells with Hoechst staining. (The picture was taken by Masur in January 2007 and placed in the public domain. Accessed January 2021 from https://commons. wikimedia.org/wiki/ File:HeLa_cells_stained_ with_Hoechst_33258.jpg)

molecules outside the cells (extracellular matrix) or inside the cells (mostly cyto-skeletons, which will be discussed in the following section). Specific membrane proteins play critical roles in cell–cell or cell–surface adhesion, which will be discussed later in this chapter.

Cytoskeleton means the bone (= skeleton) in a cell (= cyto). It consists of three components: (1) actin filaments, (2) microtubules, and (3) intermediate filaments. *Actin filaments* are two-stranded helical polymers made from the protein *actin*. They have flexible structures of diameter 5–9 nm. As shown in Fig. 2.3, they are most highly organized on the periphery of a cell, while the actin fibers that cross a cell (stress fibers) can also be found. Actins are initially formed as G-actin, where "G" stands for "globular," that is, rounded shape. Polymerization of G-actins leads to F-actin, where "F" stands for "filament." This polymerization (growth) proceeds primarily in one direction, with a fast-growing "+" end. This polymerization rate varies by the cell activity, that is, whether the cells are actively proliferating (dividing), moving, or quiescent (neither proliferating nor moving). For the cells actively proliferating and/or moving, the mean filament lifetime is relatively short (8 min), with a small fraction (40%) of actins polymerized. For the quiescent cells, that is, anchorage-dependent cells occupying most of the surface (= confluent monolayer, discussed in the next section), the mean filament lifetime is quite long (40 min) with a large fraction (70%) of actins polymerized. Actin filaments play a crucial role in cell adhesion to a surface for anchorage-dependent cells.

Microtubules are hollow cylinders comprising the protein *tubulin*. Their outer diameter is 25 nm, much larger than that of actin filaments. They are also much more rigid and straight than actin filaments due to their cylindrical, tubular structure. Microtubules are usually attached to a single microtubule-organizing center, as shown in Fig. 2.3. Like actin filaments, microtubules are polymerized from tubulins, and this polymerization proceeds in one direction, with a fast-growing "+" end. Microtubules play vital roles in the structure of cilia and flagella of cells, which provides rigidity and assists in generating cellular motion and cell division through forming mitotic spindle.

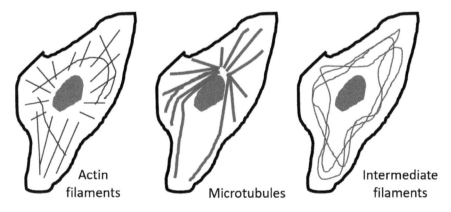

Fig. 2.3 Actin filaments (left), microtubules (middle), and intermediate filaments (right)

Intermediate filaments are rope-like fibers made from a heterogeneous family of proteins. They are quite long, with a diameter of around 10 nm. They form a mesh-work across the cell, providing mechanical strength.

2.3   Focal Adhesion

As mentioned, actin filaments are the most critical cytoskeleton in cell culture, involved in cell adhesion to a surface. For the cells in mammals and humans, that is, cells in vivo, such surface is *extracellular matrix* (*ECM*). ECM provides a surface where anchorage-dependent cells can be anchored to and give a specific structure built into tissue and ultimately an organ. In cell culture, it is essential to provide an artificial surface that mimics such ECM. The most desirable cell adhesion to a sur-face is *focal adhesion* (for anchorage-dependent cells), and its schematic is shown in Fig. 2.4. In focal adhesion, certain proteins in ECM or a specific structure and character of an artificial surface can be recognized by a membrane protein *integrin*. Such binding to integrin triggers adaptor proteins' binding within the cell, including actinin, talin, tensin, paxillin, and vinculin. *Vinculin* is the essential protein that can connect integrin and actin filaments. While other adaptor proteins may or may not be found at the focal adhesion site, you can always find vinculin. Identification of vinculin has been used as a preferred method of confirming focal adhesion of anchorage-dependent cells. In this manner, a continuous link is formed from a sur-face, cell membrane (integrin), to the cytoskeleton (actin filaments), providing quite

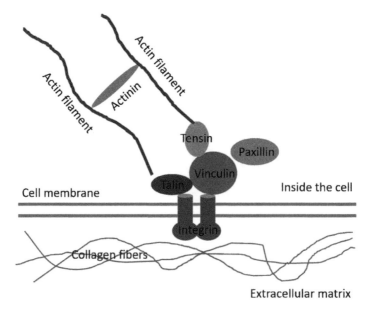

Fig. 2.4   Focal adhesion of an anchorage-dependent cell to a surface

a tight adhesion of cells to a surface. Focal adhesion will be further discussed in Chap. 7.

2.4   Cell Classification: Anchorage-Dependent Versus Anchorage-Independent Cells

At this point, you will probably have a good idea of the difference between anchorage-dependent and anchorage-independent cells. *Anchorage-dependent cells* require a surface they can anchor. The surface is typically extracellular matrix (ECM) in vivo, while an artificial surface can also be used for the cell culture in vitro. Anchorage-dependent cells cannot be stacked in multilayers on a surface, that is, they can only form a monolayer. When a surface is saturated with anchorage-dependent cells in a monolayer (caused by *contact inhibition*), such condition is referred to as *confluency*, for example, 80% confluency represents 80% of the surface is covered by anchorage-dependent cells.

On the other hand, *anchorage-independent cells* do not require a surface for their proliferation and metabolism and can be cultured in a suspension. The best examples are blood cells, including red blood cells (RBCs) and white blood cells (WBCs). As you can imagine, it will be quite problematic if blood cells are anchored to a surface (in this case, the wall of blood vessels) – they should be able to move along the blood flow to circulate throughout the body.

Anchorage-dependent cells can lose their anchorage-dependency and become anchorage-independent. A famous example is cancer cells – where the normal cells become "crazy," dividing indefinitely (no Hayflick limit), not fulfilling their assigned duties, consuming a large number of nutrients and oxygen, and converting the nearby cells into cancer cells. Sounds familiar? Yes, zombies! All those zombie movies are inspired by these crazy cancer cells, although such mutations do not occur that quickly. As cancer cells are not doing their original duties and forgetting who they were, they gradually forget their anchorage-dependency (Fig. 2.5).

2.5   Cell Classification: Normal Versus Immortalized Cells

Cells can also be classified based on their growth behavior, that is, normal versus immortalized. *Normal cells* can proliferate (divide) for a limited number of times (typically limited by Hayflick limit; discussed later in Sec. 10.2) and are highly differentiated, that is, they perform specific functions. They usually require a variety of growth factors and other components in the growth media, that is, a serum is preferred as growth media, which will be discussed in the later section of this chapter.

On the other hand, *immortalized cells* can proliferate (divide) for an indefinite number of times, thus immortal. They are sometimes referred to as a more generic

Fig. 2.5 Top: Chinese hamster ovary (CHO) cells representing anchorage-dependent cells. The picture was taken by Alcibiades in February 2006 and placed in the public domain (Accessed January 2021 from https://commons.wikimedia.org/wiki/File:Cho_cells_adherend2.jpg). Bottom: human neutrophils (more giant cells with multiple nuclei within cells; a part of human white blood cells) shown together with human red blood cells (smaller, donut-like cells), representing anchorage-independent cells. The picture was taken by Dr. Graham Bears in August 2012 and placed in the public domain. (Accessed January 2021 from https://commons.wikimedia.org/wiki/File:Neutrophils.jpg)

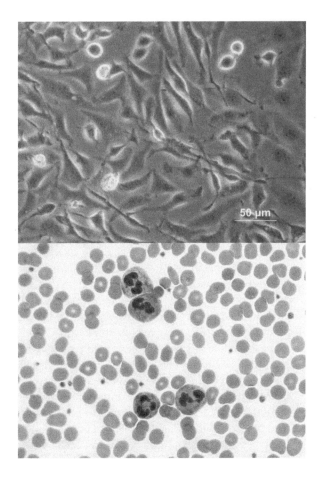

term of *transformed cells*, although these two terms' specific meanings may slightly differ. Cancer cells are considered as spontaneously immortalized cells. If the immortalized cells were originally anchorage-dependent cells, they lose their anchorage-dependency in full or in part and become (partly) anchorage-independent. Once they lose their anchorage-dependency, they can grow in suspension or multi-layers without contact inhibition. As they tend to lose their differentiation, it is essential to retain some differentiation if you plan to use immortalized cells for tissue engineering applications. As they are "crazy" cells, they are not picky eaters and may not need sophisticated serum as their growth media.

As discussed earlier in this chapter, a healthy cell culture requires (1) a viable state and (2) a metabolically active state. As immortalized cells focus on the first condition of a viable state, they comprise the second condition of the metabolically active state. Therefore, their metabolism, protein production (including cell signaling), and response to environmental factors (including anchorage-dependency) can be compromised in part or in full. Besides, the rate of DNA mutation increases, and it becomes harder to fix it.

Various methods can be used to immortalize normal cells. One of the standard techniques is infecting normal *mammalian cells* (= human and mammal cells) with a virus that can cause cancer. *SV-40* (*simian vacuolating virus 40* or *simian virus 40* in short) is a good example, which is a small DNA virus found in monkeys and humans (thus "simian") and can cause cancer to them. Various mammalian cells have been infected with SV-40, causing them to partially or fully lose their serum requirement, anchorage-dependency, and thus gaining immortality.

Primary cells and cell lines are often used in place of normal versus immortalized cells. *Primary cells* are harvested from mammals or humans, and they are most likely normal cells, that is, differentiated cells. On the other hand, *cell lines* are the ones that descended from a single cell and thus have been passaged for a long time (refer to the next section of cell passaging). While both normal and immortalized cells can be used for cell lines, cell lines are most likely immortalized cells as they can proliferate (divide) for an indefinite number of times. In this sense, normal versus immortalized cells and primary cells versus cell lines are sometimes used interchangeably, although their exact meanings differ.

Both primary cells and cell lines should be stored frozen for long-term storage. Most refrigerators have a dedicated section for freezing, and exclusive freezers are also available. However, these freezers can only provide a temperature around $-20°C$ and are inappropriate for storing mammalian cells for the long term. Because of this, *deep freezers* are available, which can provide $-40\ °C$ or $-80\ °C$, respectively (Fig. 2.6, left). Of course, the $-80\ °C$ deep freezer is substantially more expensive than the $-40\ °C$ one. However, the ultimate solution is to store these cells in liquid nitrogen storage, which will provide $-196\ °C$ (the boiling temperature of liquid nitrogen) (Fig. 2.6, right).

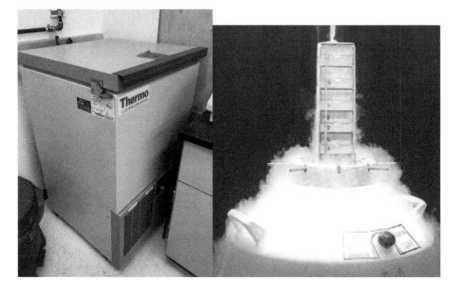

Fig. 2.6 $-40\ °C$ deep freezer (left) and liquid nitrogen tank (right) for storing mammalian cells

2.6   Cell Classification: Normal Versus Stem Cells

In the previous section, we briefly discussed differentiation. All cells in a human body retain identical genetic information, and they originate from a single type of cell (zygote = fertilized egg). In this sense, it is *totipotent* (toti = all; potent = ability) *stem cell*. You can also find stem cells in an adult human body that can differentiate into a large number of cell types but not a whole organism, called *pluripotent stem cells*, or a limited number of cell types, called *multipotent stem cells*. Examples include mesenchymal stem cells and hematopoietic stem cells. These stem cells will be discussed in Chap. 5.

Stem cells are undifferentiated and cannot perform specific functions, while normal cells are fully differentiated and perform particular operations. Stem cells can also proliferate (divide) for a small number of doublings or are close to immortal. However, stem cells are fundamentally different from immortalized cells (including cancer cells) as they are not "crazy" cells.

In tissue engineering applications, normal cells may not be ideal as they can proliferate only for a limited number of doublings and require specific growth media like a serum. The use of immortalized cells may cause several issues, as they are "crazy" cells and sometimes challenging to retain their functionality. The use of stem cells is ideal as they can proliferate for a great number of doublings and are not "crazy" cells. Differentiation can be induced later to make the cells perform specific duties. However, such differentiation is a relatively complex and delicate process, and the full and exact mechanism of differentiation is often unknown.

2.7   Maintaining Sterile Environment: Biosafety Cabinet

Cell culture can be contaminated by bacteria or yeast, which is quite a common problem. In such a case, mammal and human cells (= *mammalian cells*) compete with bacteria and yeast for nutrients. Mammalian cells may quickly lose such competition, and the entire cell culture is overrun by bacteria or yeast. Maintaining a sterile environment is mandatory to avoid such contamination. Preparation of cells and culture media and in-culture practices (feeding and passaging, which will be discussed in Chap. 3) should be conducted in a *biosafety cabinet*, also known as laminar flow hood. A typical biosafety cabinet is shown in Fig. 2.7, which is a class II biosafety cabinet.

Class I biosafety cabinet protects the laboratory and personnel, although it does not provide a sterile environment within the cabinet. Class II biosafety cabinet is the most common. It protects the laboratory and personnel (class I's function) and additionally provides a sterile environment within the cabinet. Class III biosafety cabinet protects the laboratory and personnel (class I's function), provides a sterile environment within the cabinet (class II's function), and provides additional protection for high-risk biological agents.

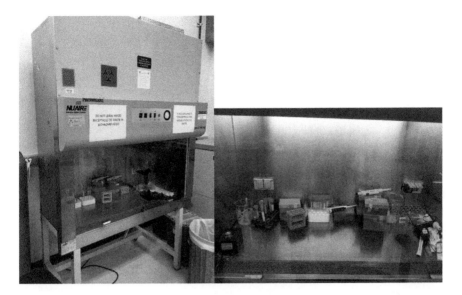

Fig. 2.7   A class II biosafety cabinet: (left) overall view and (right) inside view

The core components of a biosafety cabinet are (1) a *high-efficiency particulate air (HEPA) filter* and (2) an ultraviolet (UV) lamp. Both are typically located at the top ceiling of the inside cabinet. HEPA filter removes particulates and aerosols and must be replaced periodically. UV lamp kills bacteria, yeast, etc., that is, germicidal. Figure 2.7 (right) was taken during the active experiments, with tubes, tube racks, pipette tips, pipette tip racks, micropipettes, disposable pipettes, etc. All of them should be removed after experiments are finished, and the inside surfaces must be disinfected using bleach solution followed by ethanol. Most biosafety cabinets are equipped with power outlets connected to the AC power outlets, allowing small AC-powered equipment such as vortex mixer, rock shaker, mini-centrifuge, etc. They are also equipped with connectors that can be connected to nitrogen or compressed air tank and a vacuum pump. A vacuum is handy for removing the supernatants after centrifuging.

2.8   Maintaining Sterile Environment: Autoclave

An autoclave is another essential piece of equipment for maintaining a sterile laboratory environment. Tubes, transfer pipettes, pipette tips, glassware, etc., can also be contaminated with bacteria or yeast and must be disinfected. As many of these items are disposable, that is, for single use, a large quantity of them should be disinfected. Their smaller size and large amount make them difficult to be disinfected. An autoclave is quite useful to disinfect them altogether. As shown in Fig. 2.8, these tubes

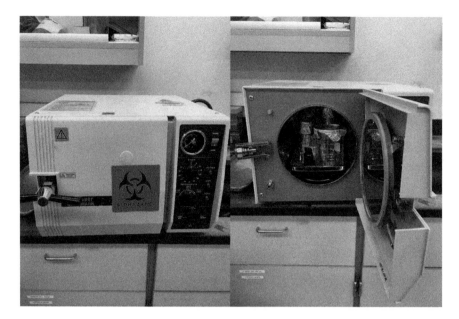

Fig. 2.8 Autoclave

and tips can be placed inside an autoclave, and high temperature (over 120 °C) and high pressure (above 15 PSI = 0.1 MPa) are applied to disinfect (= kill) any bacteria, yeast, etc.

2.9   Cell Culture: CO_2 Incubator

Cell culture can be conducted in either an incubator or a bioreactor. While a sophisticated bioreactor is almost mandatory for culturing cells in a large quantity and a continuous and automated manner, an incubator can also be conveniently used for small-scale cell culture. For most tissue engineering applications, both incubator and bioreactor are necessary, as shown in Chap. 1 and Fig. 1.2. Specifically, if many different cell culture conditions are required for optimizing cell culture parameters, an incubator becomes essential.

Incubator provides an environment that is optimum for cell culture. It typically provides a fixed temperature, for example, 37 °C (human's body temperature), optimum relative humidity (RH), for example, at 95% (preventing water evaporation), and HEPA-filtered air (providing oxygen but not particulates and aerosols). For culturing mammal or human cells, that is, mammalian cells, a specific incubator type is necessary, that is, CO_2 incubator. As its name indicates, it provides an additional environmental condition of CO_2, typically at 5%. This 5% CO_2 is the physiological condition of most mammalian tissues, allowing them to maintain appropriate pH

Fig. 2.9 Left: double-stacked CO_2 incubator with a CO_2 gas tank attached to it; right: multiple cell culture flasks within a CO_2 incubator

(when dissolved into water, CO_2 turns into bicarbonate and becomes acidic). Traditional incubators, for example, without CO_2, can be used to culture bacteria. Figure 2.9 shows a typical CO_2 incubator. In this case, two incubators are stacked. You can see a CO_2 gas tank next to it. Within the CO_2 incubator, multiple cell culture flasks are stacked, each being cultured with different cells and varying media conditions.

2.10 Cell Imaging: Fluorescence Microscope

Once cells are properly cultured, you may wish to image them to count their number, check their shape (morphology), and identify subcellular components. As cells are too small to be imaged by a conventional camera, a microscope becomes necessary. A typical microscope is shown in Fig. 2.10, where an objective lens is located underneath the sample stage. This setup is known as an *inverted microscope*, which is popular in cell imaging as it provides more room to a user. When the objective lens is located on top of the sample stage, it is known as an *upright microscope*, which is more traditional but less popular in cell imaging as it does not provide sufficient room to a user.

In cell imaging, it is quite common to stain different subcellular components with varying fluorescent dyes. It is quite challenging to figure out subcellular parts purely based on their shapes. For example, nuclei can be stained with a

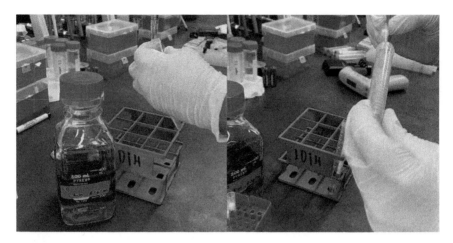

Fig. 2.10 LB (left) and lyophilized powder of *E. coli* K12 (right)

blue-fluorescent dye, actin filaments with a green-fluorescent dye, and mitochondria with a red-fluorescent dye. As it is not possible to excite all three fluorescent dyes together, a user needs to excite the fluorescent dyes one by one and acquires three different fluorescent images. These three images can be stacked together to create a single image, and of course, such an image is not something you can see through an eyepiece with your naked eye. More details on cell imaging and fluorescence staining are discussed in Chap. 4.

2.11   Laboratory Task 1: Bacterial Cell Culture

While mammalian cells are cultured in tissue engineering applications, we will learn the basics of cell culture with bacteria in this task. They are very easy to culture and less prone to contamination (bacteria themselves are contaminants to mammalian cell culture!).

Objective 1. Prepare Cells and Media
In this case, we will use lyophilized (= freeze-dried; water is evaporated at low temperature using extremely low pressure, i.e., vacuum, to avoid killing bacteria) powder of *Escherichia coli* K12, the safe strain of *E. coli*. We will use LB (lysogeny broth) as the media for *E. coli* K12.

1. Wear gloves and a lab coat. The following procedure should preferably be conducted in a biosafety cabinet.
2. Pipette 5 mL of LB into a 15-mL tube (Fig. 2.10, left).
3. Using a 1-mL pipette tip, scoop a tiny bit of lyophilized *E. coli* K12 and place it in a 15-mL tube (Fig. 2.10, right). *E. coli* powder should barely cover the pipette tip.

4. Close the lid of a tube and shake it gently using your hand (Fig. 2.11, left).

Objective 2. Culturing in an Incubator

5. Place the tube in an incubator (Fig. 2.11, right). A regular incubator is preferred. It may be possible to use a CO_2 incubator without a CO_2 supply. However, it may contaminate the CO_2 incubator and should be used with caution, and the inside cabinet should thoroughly be disinfected after the culture.

6. Maintain 37 °C and culture for 6 h or overnight (~12 h) (Fig. 2.12).

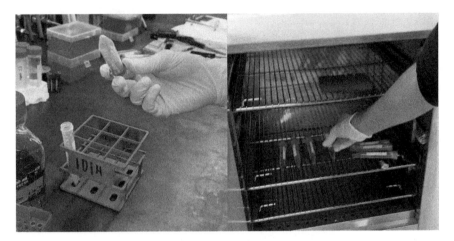

Fig. 2.11   After gentle mixing of LB and *E. coli* K12 powder (left), tubes are placed in an incubator at 37 °C (right)

Fig. 2.12   *E. coli* suspension culture over time, from 2 h (left), 4 h (middle left), 6 h (middle right), and overnight (~12 h; right). You can observe more gas formation in the early phase (= active metabolism) and more turbid solution in the later phase (= a great number of *E. coli* cells)

Objective 3. Spectrophotometric Quantification

The number of *E. coli* cells can be quantified using spectrophotometry (Fig. 2.13). As *E. coli* cells absorb and scatter incoming light, absorbance can be correlated to the *E. coli* concentration. Maximum absorbance can be observed in the visible spectrum's red color, for example, from 550 to 750 nm, any wavelength in this region can be utilized (Fig. 2.14). We will use 600 nm in this case.

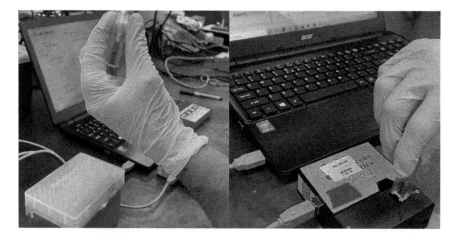

Fig. 2.13 Left: LB (as a blank) or *E. coli* culture suspension is loaded into a cuvette; right: the cuvette with the sample is placed within a spectrophotometer (a miniature spectrometer with a light source and a cuvette holder from Ocean Optics)

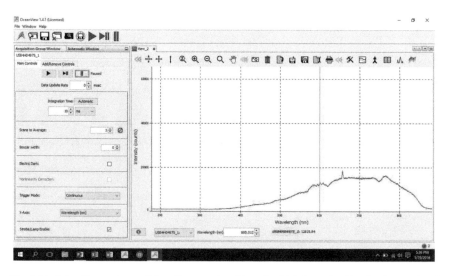

Fig. 2.14 Intensity spectrum of *E. coli* suspension. Note that high-intensity (*I*) value is correlated to smaller absorbance (*A*), as $A = -\log I / I_0$

7. Dispense 0.5 mL of LB into a disposable micro-cuvette (path length of 1 cm)
   (Fig. 2.13, left). Set this as a blank for a spectrophotometer. (You can also set
   water as a blank and subtract this LB solution's absorbance from all absorbance
   measurements.) Wipe the exterior of each cuvette with KimWipes (delicate task
   wipes), if necessary. Discard the micro-cuvette after each use. If a larger volume
   is needed for disposable cuvettes, increase the volume of LB solution and tube
   size in step 2 as needed.
8. Take 0.5 mL of *E. coli* culture at 0 h, 2 h, 4 h, 6 h, and 12 h (overnight – optional).
   If sediments (precipitates) are visible at the bottom of the tube, do not disturb
   them and draw from the supernatant. Measure absorbance (*A*) at 600 nm using a
   spectrophotometer (Fig. 2.13, right). If the spectrometer is calibrated with the
   blank solution (step 7), it will provide absorbance (*A*) values. If not, it will provide intensity (*I*) values. If this is the case, use the equation $A = -\log I / I_0$, where
   I_0 is the intensity of blank solution (in this case LB), and *I* is the intensity of your
   bacterial suspensions.
9. Plot the absorbance against time (Fig. 2.15).

The result shown in Fig. 2.15 follows the classic cell growth kinetics shown in
Fig. 2.16 quite well. The first couple of hours correspond to the exponential growth
phase. After 4 h, the growth starts to slow down and is plateauing, corresponding to
the stationary phase. While this curve was constructed from bacterial cell culture,
similar behavior can be observed with mammalian cells. More details on cell growth
kinetics will be covered in Chap. 3.

While a simple spectrophotometric measurement was used in this laboratory
exercise, plating and colony counting is a preferred bacterial quantification method.
It is more accurate, accounting only for the viable cells. However, we will not cover
this method in this chapter as it is not relevant to mammalian cell culture.

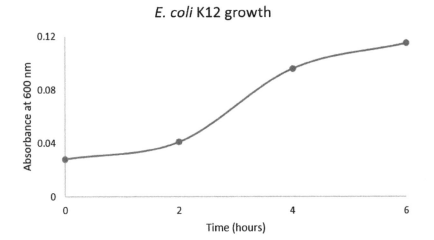

Fig. 2.15 Absorbance values at 600 nm for *E. coli* K12 culture are plotted against time

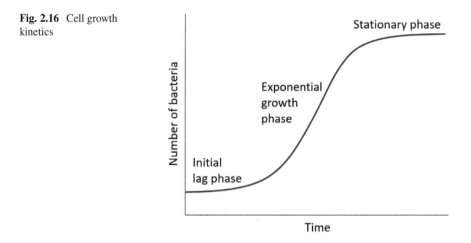

Fig. 2.16 Cell growth kinetics

2.12   Laboratory Task 2: Fluorescence Microscopic Imaging of Mammalian Cells

While we have a dedicated chapter on this topic, it will be useful to familiarize yourself with the use of a fluorescence microscope (Fig. 2.17).

10. Purchase a prepared slide of mammalian cells, for example, FluoCells Prepared Microscope Slides from Invitrogen (part of ThermoFisher Scientific).
11. Secure three different filter cubes for three fluorescent dyes. For example, cell nuclei are typically stained with DAPI (functioning as both bioreceptor for nuclei and blue-fluorescent dye). A DAPI filter cube should be installed in one slot of the filter cube holder. Filter cube can also be selected based on the excitation and emission wavelengths of the fluorescent dye. Likewise, actin filaments are typically stained with FITC-phalloidin, where phalloidin is a receptor binding to actin filaments, and FITC is a green-fluorescent dye. Mitochondria are stained with tetramethylrhodamine methyl ester (TMRM). It can bind to mitochondria and exhibit red fluorescence due to the rhodamine presence in its structure.
12. Install all three filter cubes. You can find the filter cube holder underneath the objective lens. Position the DAPI filter cube in the place so that cell nuclei can be imaged.
13. Move the sample stage to find the cells (cell nuclei as a DAPI filter cube is used) using either an eyepiece or an image shown in the attached computer. Choose an appropriate objective (10×, 20×, 30×, etc.) and adjust focus and magnification (Fig. 2.18). Save the image (DAPI-stained cell nuclei).
14. Slide the filter cube holder to position the FITC filter cube. Save the image (FITC-phalloidin-stained actin filaments).
15. Slide the filter cube one more time to position the TRITC filter cube (used for rhodamine dye). Save the image (TMRM-stained mitochondria).
16. Stack three acquired images into one (Fig. 2.19). Depending on the prepared slide, staining can vary.

Fig. 2.17 A fluorescence
microscope

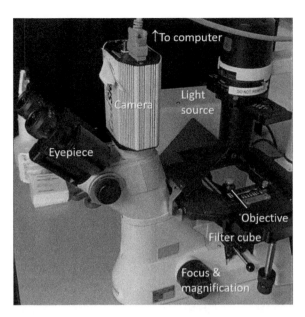

Fig. 2.18 Cell nuclei are
being focused on the
real-time images shown on
a computer connected to a
microscope

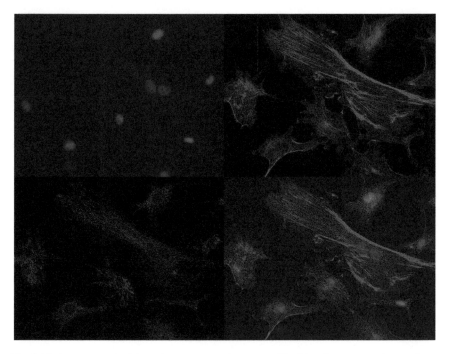

Fig. 2.19 Images acquired by a fluorescence microscope. The blue fluorescence image shows nuclei (top left), the green fluorescence image shows actin filaments (top right), the red fluorescence image shows mitochondria (bottom left), and the stacked image is finally shown (bottom right). (Reprinted with permission from Yoon, 2016. (C) 2016 Springer)

Review Questions
1. Describe three main objectives of cell culture.
2. What is a cell line, and how is it different from a primary cell?
3. What are membrane proteins, and how can they be stable within the cell membrane?
4. Describe the differences in actin filaments, microtubules, and intermediate filaments.
5. Describe the differences between anchorage-dependent and anchorage-independent cells.
6. What is focal adhesion? Why is it important for anchorage-dependent cells?
7. What is cell immortalization? How can it be done?
8. What are the negative impacts of cell immortalization?
9. How is a biosafety cabinet different from a chemical hood?
10. Why is CO_2 needed for mammalian cell culture?
11. Compare two different incubators for bacterial and mammalian cell culture.
12. Draw a typical cell growth curve. Identify each phase.

Reference

Yoon, J. Y. (2016). *Introduction to biosensors* (2nd ed.). Springer, Chapter 9. https://doi.org/10.1007/978-3-319-27413-3_9

Chapter 3
Cell Metabolism

In the previous chapter, we have learned the basics of cell culture. In this chapter, we will learn the details of cell culture – how we can feed and grow them. Here are three inquires for you:

Inquiry 1. What is cell metabolism?

How do cells "eat" nutrients and metabolize them with oxygen? And how do cells generate energy (ATPs) and produce proteins?

Inquiry 2. How do we feed cells?

More specifically, how do we prepare the media for cells? Should we replace or replenish media periodically? Why do we need to "passage" cells?

Inquiry 3. How can we determine the initial cell number (or density) for tissue engineering applications?

It can be calculated using the first-order growth model under varying cell culture restrictions.

3.1 What Is Cell Metabolism?

Cells require nutrients (and oxygen) to generate energy, produce proteins, and proliferate (divide) themselves. Nutrients include sugars, amino acids, small molecules, etc. Energy is generated in the form of *ATP* (*adenosine triphosphate*), which is the primary energy currency in cell metabolism (other forms of energy currency are also available). Cells can also produce small molecules, such as lactic acid, ammonia, water, etc., as waste or by-products. These processes, taken together, form *cell metabolism*. It can be divided into two parts: (1) *catabolic reaction*, that is, break-down of complex molecules to generate energy (e.g., ATP) while generating small molecules as waste or by-product, and (2) *anabolic reaction*, that is, biosynthesis of

© Springer Nature Switzerland AG 2022
J.-Y. Yoon, *Tissue Engineering*, https://doi.org/10.1007/978-3-030-83696-2_3

large molecules (proteins, nucleic acids, fats, cholesterol, etc.) that consumes energy (e.g., ATP).

For optimum cell metabolism, we will need to provide optimum physical conditions. As we learned in the previous chapter, we know that a CO_2 incubator provides the 37°C temperature, 95% relative humidity (RH), HEPA-filtered air (providing oxygen but not contaminants), and 5% CO_2. In addition to these physical conditions, we will need to provide optimum metabolic conditions as shown below:

- The pH of media should be maintained 7.2–7.5, that is, *physiological pH*. While buffers can maintain such optimum pH, CO_2 also affects the medium pH as the dissolved CO_2 becomes bicarbonate (HCO_3^-), making the medium acidic.
- Nutrients should be provided. The primary carbon source is glucose, while the primary nitrogen source is glutamine.
- Secondary requirements should also be provided. These include growth factors, cofactors, vitamins, and trace elements (mostly metal ions including Fe, Mg, Mn, Ca, Zn, etc.).
- Waste products should be removed, for example, lactate from glucose consumption and ammonia from glutamine consumption.

3.2   Energy Currency: ATP

As mentioned above, *ATP (adenosine triphosphate)* is the most widely used energy currency. Figure 3.1 shows the molecular structure of ATP.

ATP has three phosphate groups (P_i), which exhibit the highest level of energy. When ATP loses one phosphate group, it becomes *ADP (adenosine diphosphate)* and generates energy. It can lose one more phosphate group, becoming *AMP (adenosine monophosphate)* and generates additional energy.

$$ATP \rightarrow ADP + P_i + energy$$
$$ADP \rightarrow AMP + P_i + energy$$

Fig. 3.1 ATP (adenosine triphosphate). (The picture was taken by NEUROtiker in June 2007 and placed in the public domain. Accessed January 2021 from https://commons.wikimedia.org/wiki/File:Adenosintriphosphat_protoniert.svg)

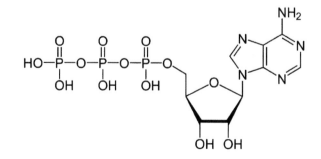

AMP can return to ATP through acquiring two phosphate groups and ADP to ATP through acquiring one phosphate group, both of which require energy. ATPs can be generated by the oxidation of fuel molecules, for example, sugar, fatty acids, and amino acids. The primary fuel molecules are glucose (the simplest form of sugar) and glutamine (the most abundant free amino acids in the mammal and human body).

There are other types of energy currency: *NADH* (*nicotinamide adenine dinucleotide*) and *FADH$_2$* (*flavin adenine dinucleotide*).

$$NADH \rightarrow NAD^+ + H^+ + energy$$
$$FADH_2 \rightarrow FADH + H^+ + energy$$

The oxidized NAD^+ and FADH can also return to the reduced forms of NADH and FADH$_2$ through gaining H^+, which require energy.

3.3   Cell Metabolism: Glycolysis and TCA Cycle

Metabolism of primary nutrients of glucose and glutamine can be classified into glycolysis (glucose metabolism), glutaminolysis (glutamine metabolism), and their later common TCA cycle pathway shown in Fig. 3.2.

In *glycolysis*, the input material is glucose, the simplest form of sugars, and its main function is to generate ATPs. It is sometimes referred to as the EMP pathway, named after the Embden–Meyerhof–Parnas. While the whole pathway shown in Fig. 3.2 is quite complicated involving multiple steps, the whole reaction can be simplified to

$$Glucose + 2ADP + 2P_i + 2NAD^+ \rightarrow 2Pyruvate + 2ATP + 2NADH + 2H^+ + 2H_2O$$

Consumption of one glucose molecule will generate two ATPs and two NADHs with the final product of *pyruvate*. The entire reaction occurs in the *cytosol*, that is, the aqueous component of a cell. It will also generate the waste product of *lactate*, as shown in Fig. 3.2. Note that the process does not require oxygen, that is, *anaerobic reaction*.

Pyruvate will further be converted into *acetyl-CoA* (*acetyl coenzyme A*), generating one additional NADH. This reaction is referred to as *pyruvate decarboxylation*.

$$2Pyruvate + 2NAD^+ + 2CoA \rightarrow 2Acetyl\ CoA + 2NADH + 2CO_2 + 2H^+$$

Acetyl-CoA then enters the *TCA cycle* (*tricarboxylic acid cycle*), known as the *citric acid cycle*, as illustrated in Fig. 3.2. It is a cyclic reaction, again involving multiple steps. Both pyruvate decarboxylation and TCA cycle occur in the *mitochondrion* (plural = mitochondria; a small organelle found in most eukaryotic cells, including

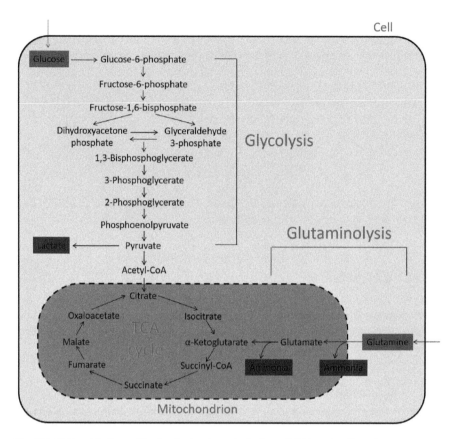

Fig. 3.2 Cell metabolism with glycolysis, glutaminolysis, and TCA cycle. Green boxes represent primary nutrients (glucose and glutamine), and red boxes wastes (lactate and ammonia). A specific enzyme catalyzes each step. Production and consumption of ATP, NADH, FADH$_2$, and CO$_2$ in each step are not shown

mammal and human cells). Overall, one acetyl-CoA molecule consumption will generate one ATP (or GTP = guanosine triphosphate), three NADH, and one FADH$_2$. As consumption of one glucose will generate two pyruvates and two acetyl-CoA molecules, it will generate two ATPs (or GTPs), six NADH, and two FADH$_2$. Therefore, the simplified overall reaction is

$$2\text{Acetyl } \text{CoA} + 2\text{ADP}\left(\text{orGDP}\right) + 2\text{P}_i + 6\text{NAD}^+ + 2\text{FAD} + 4\text{H}_2\text{O}$$
$$\rightarrow 2\text{CoA } \text{SH} + 2\text{ATP}\left(\text{orGTP}\right) + 6\text{NADH} + 6\text{H}^+ + 2\text{FADH}_2 + 4\text{CO}_2$$

Again, the process does not require oxygen, that is, anaerobic reaction. From glycolysis to TCA cycle, consumption of one glucose molecule will generate the energy currency shown in Table 3.1.

Table 3.1 Energy currency generated by glycolysis and the TCA cycle from consuming one glucose molecule

	ATP	NADH	FADH$_2$
Glycolysis	2	2	0
Pyruvate decarboxylation	0	2	0
TCA cycle	2	6	2
Total (anaerobic)	4	10	2
Oxidative phosphorylation		25 ATPs added	3 ATPs added
Total (aerobic)	32	0	0

Note that both glycolysis and TCA cycle are inherently anaerobic, that is, do not require oxygen. However, when oxygen is added to the TCA cycle, it will convert NADH and H$^+$ back to the oxidized state NAD$^+$:

$$NADH + \tfrac{1}{2}O_2 + H^+ \rightarrow H_2O + NAD^+$$

Electrons generated from the reaction are used to power the formation of ATP.

$$ADP + Pi + H^+ \rightarrow ATP + H_2O$$

This process is referred to as *oxidative phosphorylation* as the addition of oxygen adds phosphate groups to form ATPs. Each NADH yields about 2.5 ATPs. Likewise, each FADH$_2$ yields about 1.5 ATPs. As shown in Table 3.1, consumption of one glucose molecule generates 10 NADH and 2 FADH$_2$. Additional 25 ATPs (from 10 NADH) and 3 ATPs (from 2 FADH$_2$) can be generated through oxidative phosphorylation. Therefore, the total number of ATPs generated by glycolysis and TCA cycle with oxidative phosphorylation becomes 4 + 25 + 3 = 32.

You can notice the significant difference in the ATP numbers with versus without oxygen (32 vs. 4), indicating the importance of providing an adequate amount of oxygen to the cell culture in the form of *dissolved oxygen (DO)*. However, in practical cell culture, this "ideal" number of ATP production is rather difficult to achieve.

3.4 Glucose Transport and Uptake

Glucose molecules must pass through a cell membrane for cells to metabolize them. Unfortunately, the cell membrane is impermeable to glucose. Therefore, a specific membrane protein – *glucose transporter* – is necessary to transport glucose across a cell membrane actively. Usually, glucose will be transported from the high concentration area (typically the outside of a cell) to the low concentration area (typically inside a cell). However, such transport is not a passive molecular diffusion but rather an active delivery mechanism. Because of this, it requires energy, that is, it consumes ATP. However, this is not a problem as consumption of one glucose molecule can generate as much as 32 ATPs with oxidative phosphorylation. Glucose consumption inside a cell is quite a rapid process, especially with oxygen (oxidative phosphorylation in the TCA cycle). The glucose concentration inside a cell is typically very low.

Multiple glucose transporters (they are all membrane proteins) have been identified, and they share similar peptide sequences and, accordingly, similar structures. These molecules not only "transport" glucose across the cell membrane but also regulate the rate of uptake to maintain the appropriate intracellular concentration of glucose. Among these, *GLUT1* is the most well-known and primary glucose transporter in most mammalian cells (Fig. 3.3).

When cell metabolism is deregulated (e.g., in immortalized cells and cancer cells), the glucose uptake rate increases. Such abnormal uptake leads to incomplete glycolysis and increased lactate (waste) production. The incomplete glycolysis subsequently leads to the minimum entry into the TCA cycle, substantially decreasing the ATP production even under oxidative phosphorylation.

Feeding cells with high glucose concentration generate similar results as the high glucose concentration outside a cell increases the uptake rate. Glucose is more efficiently consumed (i.e., generating the maximum number of ATPs and minimum lactate molecules) when glucose concentration is lower. High glucose feeding leads to (1) low conversion of glucose to pyruvate, (2) more lactate production, and (3) minimal entry into the TCA cycle. Inadequate glucose feeding leads to, on the other hand, (1) high conversion of glucose to pyruvate, (2) less lactate production, and (3) more entry into the TCA cycle.

Unfortunately, mammalian cells in culture are highly deregulated (whether the cells are immortalized or not). Therefore, they tend to consume more glucose at a higher uptake rate than needed for their maintenance and growth. As a result, the above issues – more lactate production and smaller ATP production – become problematic in practicing cell culture. These inferior performances can be attributed to multiple factors, including (1) immortalization, (2) selection of a cell line optimized for one specific purpose, and (3) difference in cell culture environment from that of

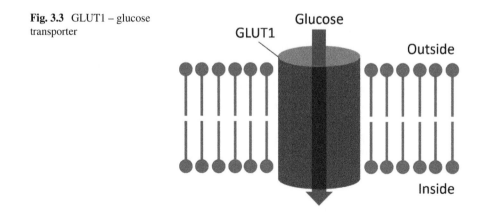

Fig. 3.3 GLUT1 – glucose transporter

tissue, etc. Similar issues can be found in glutaminolysis, which will be discussed in the following section.

3.5 Cell Metabolism: Glutaminolysis and TCA Cycle

As mentioned previously, glutamine is the most common form of amino acids found in mammal and human bodies and can be consumed to generate ATPs. *Glutaminolysis* (= glutamine metabolism) is substantially simpler than glycolysis, as shown in Fig. 3.2. First, glutamine is converted to glutamate with the enzyme phosphate-activated glutaminase (PAG) at the mitochondrial membrane. One ammonium molecule is generated as a waste product. The second reaction will lead to the TCA cycle, generating one more ammonium molecule as a waste product. Note that the entry point to the TCA cycle is different from that of glycolysis. As a result, glutaminolysis itself does not generate ATPs. However, upon entry to the TCA cycle and with oxidative phosphorylation, it can generate many ATPs, although still inferior to that of glucose consumption.

3.6 Glutamine Transport and Uptake

Like glucose, glutamine also needs to be actively transported across a cell membrane. Also, like glucose transporters, they are several types of *glutamine transporters*. However, the major difference is that glutamine transporters are less specific than glucose transporters as it can transport all kinds of amino acids while glucose transporter transports only glucose.

Like glucose transport, immortalized cells and cancer cells can uptake and metabolize glutamine at higher rates. This abnormal uptake leads to increased ammonia (waste) production, incomplete glutaminolysis, and minimal entry into the TCA cycle (and decreased ATP production). Likewise, feeding cells with high glutamine concentration generates similar results, leading to more ammonia production and minimal entry into the TCA cycle (and decreased ATP production).

Again, mammalian cells in culture are highly deregulated. They tend to consume more glutamine at a higher uptake rate, leading to more ammonia production and smaller ATP production per glutamine consumption.

There is a strong correlation between glucose uptake and glutamine uptake. When the glucose concentration is higher than the required minimum (typically 1 mM), glutamine uptake is not affected by glucose concentration and increases with glutamine concentration. Similarly, when the glutamine concentration is higher than the required minimum (typically 0.2 mM), glucose uptake is not affected by glutamine concentration and increases with glucose concentration. In other words, they are not affecting each other's uptake rate.

When the glucose concentration is lower than the minimum (< 1 mM), glutamine uptake increases significantly to find an alternative supply of nutrients. Likewise, when the glutamine concentration is lower than the minimum (< 0.2 mM), glucose uptake increases significantly again to find an alternative nutrient supply.

For quantifying the extents of cell metabolism, including both consumption and production, *metabolic rate* is commonly used. It is usually represented as q, in the unit of g/cell-h, representing the mass (in g) of substrate (e.g., glucose or glutamine) consumption or production (ATP, lactate, ammonia, etc.) per single cell per hour. The unit of mol/cell-hr is also used. The following questions utilize these metabolic rates to quantify the effects of varying culture conditions.

Question 3.1 When glucose concentration increases (high glucose feeding), determine whether the following metabolic rates (q) would increase, decrease, or stay the same. Assume glutamine concentration is unchanged.

– $q_{glucose}$ (glucose consumption rate): increases as the glucose concentration increases
– $q_{glutamine}$ (glutamine consumption rate): stays the same if the glucose concentration is higher than minimum (glucose concentration does not affect glutamine consumption rate)
– $q_{lactate}$ (lactate production rate): increases as the glucose consumption rate increases
– $q_{ammonia}$ (ammonia production rate): stays the same if the glutamine consumption rate is unchanged

Question 3.2 Repeat Question 3.1:

– When the glucose concentration decreases lower than the required minimum (< 1 mM)
– When the glutamine concentration increases (high glutamine feeding)
– When the glutamine concentration decreases lower than the required minimum (< 0.2 mM)

3.7   Role of Oxygen in Cell Metabolism

As explained previously, oxygen dramatically increases the amount of ATP production in the TCA cycle. Oxygen must be dissolved in water for cells to utilize. Unfortunately, the solubility of oxygen in water and subsequently culture media is low – about 0.21 mM, which is equal to 0.21×10^{-3} mol/L $\times$ 32 g/mol (the molecular weight of O_2) $= 6.7 \times 10^{-3}$ g/L $= 6.7$ mg/L. The typical oxygen metabolic rate q_{O2} (oxygen consumption rate) of mammalian cells is about 10^{-8} mg/cell-h. This metabolic rate does not change much when dissolved oxygen concentration is between 10 and 100% of its saturated concentration (100% = 0.21 mM). When the oxygen saturation drops below 10%, the cell growth rate declines as ATP production through oxidative phosphorylation is compromised. When the oxygen saturation drops below 0.5%, which is an emergency, cells strive to metabolize as much glucose and glutamine as possible to generate ATPs without oxidative phosphorylation – that is, anaerobic metabolism, even though it can generate only four ATPs from a single glucose molecule (vs. the maximum of 32). Lactate and ammonia production will increase, generating quite a toxic environment for cells.

As explained in previous sections, deregulated metabolism, including high glucose feeding, high glutamine feeding, and low oxygen saturation, leads to increased lactate or ammonia production, that is, waste production. Mammalian cell cultures, especially with immortalized cells, are highly deregulated. As such, they produce more lactate or ammonia than they do in vivo. While both lactate and ammonia are toxic, ammonia is much more toxic than lactate. Ammonia is considered toxic with a concentration higher than 2 mM, while the toxic level of lactate is much higher, 20 mM (thus safer). Because of this, glucose is preferred over glutamine at higher oxygen levels. Glutamine metabolism is augmented when the oxygen level falls below a certain level.

Question 3.3 When the oxygen saturation increases to 100%, determine whether the following metabolic rates would increase, decrease, or stay the same.

- $q_{glucose}$ (glucose consumption rate): increases as glucose is the preferred substrate
- $q_{glutamine}$ (glutamine consumption rate): stays the same as glutamine is not preferred
- $q_{lactate}$ (lactate production rate): increases as the glucose consumption rate increases
- $q_{ammonia}$ (ammonia production rate): stays the same as glutamine consumption rate is unchanged

Question 3.4 Repeat Question 3.3 when the oxygen saturation decreases to 10%.

Question 3.5 What are the impacts (increase, decrease, or no effect) of low oxygen level (<10%, i.e., near anaerobic condition) for the following?

A. ATP production (hint: consider oxidative phosphorylation)
B. Waste (lactate and ammonia) production (hint: cells still need a certain number of ATPs for their survival)
C. Ratio of glucose over glutamine consumption $q_{glucose}/q_{glutamine}$ (hint: while cell generally prefers glucose over glutamine, what will happen under an emergency?)

A variety of metabolic engineering methods can reduce lactate and ammonia production. For example, the use of lactate dehydrogenase (LDH) can convert lactate back to pyruvate. The use of glutamate in place of glutamine can also reduce the number of ammonia production from two to one, as shown in Fig. 3.2, at the cost of a higher price.

3.8   Culture Media

As we learned in the previous sections, we need to provide the followings to the media of cell culture:

- Sugars, preferably glucose. The minimum required concentration is about 1 mM. Typical glucose concentration is about 20 mM in media.
- Glutamine (optional). The minimum required concentration is about 0.2 mM. Typical glutamine concentration is about 2 mM in media.
- Amino acids. Required for protein synthesis.
- Buffers to maintain proper pH, typically at 7.4. The acceptable range of physiological pH is from 7.2 to 7.5. A pH indicating dye is sometimes added to the media to monitor its pH. *Phenol red* is frequently used in cell culture as its color is yellow under pH < 6.8 and red (bright pink) under pH > 8.2.

These are just bare minimum requirements. To maintain healthy cell culture, you will additionally need growth factors, cofactors, vitamins, hormones, and minerals. Cells in vivo (i.e., in mammal and human bodies) receive all these supplies mainly from the blood. While you may be tempted to use whole blood as cell culture media, it is not practical as it contains blood cells (you do not want to mix your cells with blood cells) and clotting protein (fibrinogen; it will make your blood coagulated). Suppose you let the whole blood to coagulate by itself and centrifuge it. In that case, both blood cells and blood clots (polymerized and cross-linked from clotting proteins) will precipitate out from blood, leaving the supernatant solution referred to as *blood serum* or simply *serum*. And you can use this serum as the media for your cell culture. A serum is typically diluted to 5–10% v/v for cell culture. The apparent advantage of serum as culture media is that it has everything needed for cell growth and metabolism, including growth factors, cofactors, vitamins, hormones, and minerals, in addition to glucose, glutamine, amino acids, and buffers.

However, a serum is not perfect for cell culture. There are batch-to-batch variations, depending on the mammals used, the time of collection, nutrient conditions, etc. Also, serum causes foaming when mixed at a high rate in a bioreactor, mainly due to the presence of specific serum proteins (e.g., albumin). These proteins make the later-stage purification process difficult when cell culture is used for a particular production of protein. There also exists a potential risk of virus infection as mammals are used. Most importantly, a serum is expensive compared to the other defined media described below.

Because of these shortcomings, many companies provide *defined media* that were mixed with growth factors, cofactors, vitamins, hormones, and minerals, as well as glucose, glutamine, amino acids, buffers, and/or pH indicator dye, all at fixed concentrations (or with defined "recipe"). A partial list of well-known defined media is shown below:

- *DMEM: Dulbecco's modified Eagle's medium*
- *IMDM: Iscove's modified Dulbecco's medium*
- *RPMI 1640: Roswell Park Memorial Institute 1640 medium*

All these defined media are sterilized through filtration. There exist many variations for each defined media. For example, as many as 36 different variations of DMEM are available from the Sigma-Aldrich website at the time of writing. The "recipe" of one such DMEM is shown in Table 3.2. You can notice the presence of sodium bicarbonate – note that CO_2 becomes bicarbonate (HCO_3^-) when it is dissolved in

Table 3.2 Composition of DMEM

Category	Component	Concentration (g/L)
Inorganic salts	Calcium chloride	0.2
	Ferric nitrate · $9H_2O$	0.0001
	Magnesium sulfate (anhydrous)	0.09767
	Potassium chloride	0.4
	Sodium bicarbonate	3.7
	Sodium chloride	6.4
	Sodium phosphate monobasic (anhydrous)	0.109
Amino acids	L-Arginine · HCl	0.084
	Glycine	0.03
	L-Histidine· HCl · H_2O	0.042
	L-Isoleucine	0.105
	L-Leucine	0.105
	L-Lysine · HCl	1.46
	L-Phenylalanine	0.066
	L-Serine	0.042
	L-Threonine	0.095
	L-Tryptophan	0.016
	L-Tyrosine · 2Na · $2H_2O$	0.12037
	L-Valine	0.094
Vitamins	Choline chloride	0.004
	Folic acid	0.004
	myo-Inositol	0.0072
	Niacinamide	0.004
	D-Pantothenic acid (hemicalcium)	0.004
	Pyridoxine · HCl	0.004
	Robiflavin	0.0004
	Thiamine · HCl	0.004
Nutrients and others	D-Glucose	4.5
	Pyruvic acid · Na	0.11
	L-Glutamine (optional)	0.584
pH indicator dye	Phenol red · Na	0.0159

Retrieved from Sigma-Aldrich (2020), catalog number D0422: https://www.sigmaaldrich.com/content/dam/sigma-aldrich/docs/Sigma/Formulation/d0422for.pdf water. You can also see that glutamine is offered as optional. A pH indicator dye, phenol red, is also included to monitor the medium pH.

Advantages of defined media include (1) no batch-to-batch variation, (2) no use of mammals (except for the possible mammal use for producing individual components), (3) no viral contamination from mammals, and (4) low to no proteins, leading to no foaming and easier later purification process.

3.9   Cell Feeding

While the media will be provided at the beginning of the cell culture, many components, especially nutrients, will be consumed over time, and wastes will be produced. Therefore, cells need to be fed with new media, typically every 2–3 days. This process is called *cell feeding*.

3.10   Cell Passaging

As cells continue to proliferate (divide) in culture, they grow to fill the available surface area (for anchorage-dependent cells) or volume (for anchorage-independent cells). Such proliferation can generate multiple issues, including (1) nutrient depletion in the growth media, (2) accumulation of dead cells, (3) *contact inhibition* for anchorage-dependent cells, and (4) occasionally unwanted cellular differentiation. Both contact inhibition (for anchorage-dependent cells) and high cell density (for anchorage-independent cells) can lead the cells to *senescence*, that is, cells stop proliferating (dividing).

To prevent these complications, you should periodically split the cells into a new vessel, called *cell passaging* (Fig. 3.4). For anchorage-independent cells (suspension culture), cells can easily be passaged by transferring a small amount of cell culture into a larger volume of fresh media in a new flask (or bioreactor). However, for anchorage-dependent cells (adherent culture), cells first need to be detached from the surface. This detachment is commonly done with *trypsin-EDTA*, where trypsin is an enzyme (a protein) that can help digest proteins. However, other enzyme mixes are also available for cell passaging. A small number of detached cells can then be seed into new media in a new culture flask.

Cell passaging is mandatory when the anchorage-dependent cells reach confluency or the anchorage-independent cells reach the maximum allowable cell density. Without it, cells will go senescence and stop proliferating.

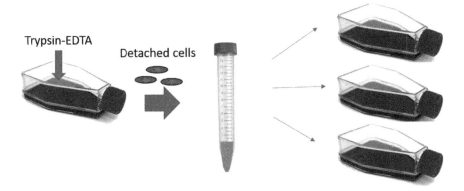

Fig. 3.4 Schematic of cell passaging

3.11   Cell Growth Kinetics: First-Order Growth Model

As we have briefly learned from laboratory task 1 of the previous chapter, the initial growth of bacterial and mammalian cells can be characterized by the exponential increase. It can be mathematically represented in the following equation:

$$\frac{dX}{dt} = \mu X$$

where X is *cell density* (cells/mL), t is time (typically in hr), and μ is *specific growth rate* (typically in the unit of h^{-1}). This model is referred to as the *first-order growth model*. It is a simple first-order differential equation. Sending X to the left side and dt to the right yields

$$\frac{dX}{X} = \mu \, dt$$

Integrating from the initial condition of $t = 0$ h and $X = X_0$ (initial cell density) to the final state at t h and X cell density

$$\int_{X_0}^{X} \frac{1}{X} dX = \int_{0}^{t} \mu dt$$
$$\ln X - \ln X_0 = \mu (t - 0)$$
$$\ln \frac{X}{X_0} = \mu t$$

In cell culture, *doubling time* t_d is often preferred over specific growth rate μ. It is defined as the time (typically in h) for given cells to double their number. It can be as short as a few hours (e.g., bacterial cells) and up to a few months (e.g., liver cells).

Many mammalian cells used for cell culture have the doubling time from a day to several days. At time t_d, X becomes $2X_0$, and plugging these into the above equation yields

$$\ln \frac{2X_0}{X_0} = \ln 2 = \mu t_d$$

$$t_d = \frac{\ln 2}{\mu}$$

The first-order growth model is essential in determining the initial cell density X_0 with various tissue engineering applications' varying design constrictions. For example, if you need a 100-mL tissue-engineered transplant or a system with the final cell density of 10^8 cells/mL, you will need to generate 10^8 cells/mL $\times$ 100 mL = 10^{10} cells. Let us assume the doubling time of 2 days (= 48 h). If you can only culture for a specific duration of time, for example, 7 days, or the cells can divide only up to 10 generations, you must start with enough cells to meet these demands. Here are a couple of example questions.

Question 3.6 You are constructing a tissue-engineered device in vitro. Here are the design parameters:

Volume = 100 mL.
Only one type of mammalian cells.
Cell density = 10^8 cells/mL.
Doubling time of the given mammalian cell = 12 h.
Assuming the first-order growth model, calculate the mammalian cells' initial number to complete the device (transplant) in one week.

Solution. Converting doubling time (12 h) to specific growth rate

$$\mu = \frac{\ln 2}{t_d} = \frac{\ln 2}{12\,h} = 0.0578\,h^{-1}$$

Final condition: $X = 10^8$ cells/mL, $t = 7 \times 24 = 168$ h. Plugging into the above equation yields

$$\ln \frac{10^8 \text{ cells / mL}}{X_0} = 0.0578\,h^{-1} \times 168\,h = 9.70$$

$$\exp(9.70) = \frac{10^8 \text{ cells / mL}}{X_0}$$

$$X_0 = 6.1 \times 10^3 \text{ cells / mL}$$

Multiplying this X_0 with 100 mL volume yields

$$\left(6.1 \times 10^3 \text{ cells / mL}\right) \times \left(100\text{mL}\right) = 6.1 \times 10^5 \text{ cells.}$$

Question 3.7 Repeat Question 3.6 with the following new restriction: these cells can divide only up to 10 generations (doublings). There is no restriction on the culture time.

Solution. As doubling time is 12 h, 10 doublings should take 120 h.

$$\ln \frac{10^8 \text{ cells} / \text{mL}}{X_0} = 0.0578 \, \text{h}^{-1} \times 120 \, \text{h}$$

The rest will be similar.

3.12   Laboratory Task 1: Media Preparation and Cell Feeding

We will grow mammalian cells in a T-25 tissue culture flask under a CO_2 incubator and practice feeding in this task. Tissue culture flasks are available for anchorage-dependent cells (also referred to as *adherent cells*) or the ones for anchorage-independent cells (also referred to as *spheroid cells*). We will use the ones for anchorage-dependent cells in this demonstration. Their surfaces are specially treated to facilitate cellular adhesion (focal adhesion), whose technology is often proprietary. By contrast, the tissue culture flasks for anchorage-independent cells are not treated, typically bare plastic surfaces. *T-25 flask* has a growth area of 25 cm² and a maximum volume of 4 mL. *T-75 flask* is also widely used, where its growth area is 75 cm² and a full volume of 12 mL. As shown in Fig. 3.5, the tissue culture flask has an angled neck to provide easy access for a pipette during the culture.

Objective 1. Cell Culture
1. Wear gloves. Spray ethanol (70%) on hands and forearms.
2. Cells: rat vascular endothelial cells (RVECs). While many different mammalian cells can be used, we will use RVEC as an example. Vascular endothelial cells are the innermost cells in the blood vessel, directly facing the blood. They adhere to the connective tissue of the blood vessel and thus are anchorage-dependent. Let us use 1 mL of cell suspension at the concentration of 1×10^6 cells/mL, which is relatively high. Lower initial cell concentration can be used while it will take a longer time to culture.

Fig. 3.5 T-25 tissue culture flask

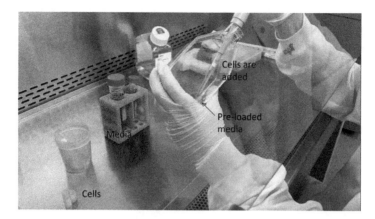

Fig. 3.6   Adding media and cells to a T-25 flask

3. Medium: DMEM. Place the media in a 37 °C water bath for at least 15 min.
4. Serum supplement: 10% v/v fetal bovine serum (FBS) is added, containing serum proteins and growth factors (lacking in DMEM).
5. Antifungal (optional): 0.2% v/v of 250 µg/mL amphotericin B is added to eliminate fungal infections.
6. Antibiotic (optional): 0.1% v/v of 50 mg/mL gentamicin sulfate is added to eliminate bacterial infections.
7. All the above are added to a T-25 cell culture flask in a biosafety cabinet. Let us use the total volume of 4 mL (1 mL cell suspension and 3 mL medium) (Fig. 3.6).
8. Place the T-25 flask in a CO_2 incubator under 37 °C and 5% CO_2.

Objective 2. Cell Imaging
9. Take the T-25 flask out from a CO_2 incubator after overnight culture. Place it on a light microscope. No fluorescence is used (Fig. 3.7).
10. Take a photograph of cells with 10× and 40× objectives.
11. Identify the morphology of cells – do they spread/stretch over the surface (ideal behavior for anchorage-dependent cells) or show somewhat spherical/rounded morphology (not ideal)?
12. Try to quantify the area covered by cells to evaluate confluency = % surface area covered by cells. ImageJ (freeware – available from https://imagej.nih. gov/ij/) is quite commonly and frequently used in microscopic imaging, and it offers a tool to evaluate confluency.

Figure 3.8 shows an example image from a T-25 flask. While a healthy, well-adhering (spreading/stretching) cell can be identified, you can also see an unhealthy cell (rounded) and even a dead cell. And the total number of cells in the image is relatively low, and the confluency is quite low. This low number is expected as proper cell feeding and cell passaging (which will be exercised in the next laboratory task) have not been implemented.

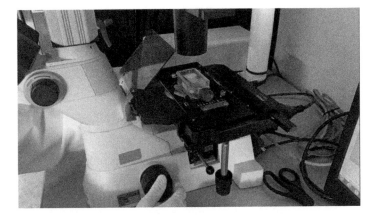

Fig. 3.7 Imaging the T-25 flask on a microscope

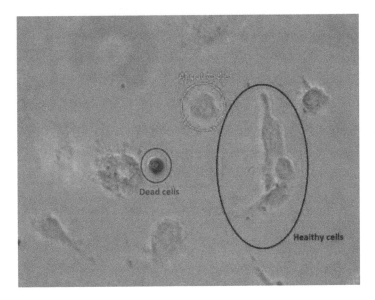

Fig. 3.8 Bright-field microscopic image of anchorage-dependent cells on a T-25 flask

Objective 3. Cell Feeding
Cell feeding is a process of replenishing the culture medium with a fresh one (Fig. 3.9).

13. Place *DPBS (Dulbecco's phosphate buffered saline)* in a 37 °C water bath for at least 15 minutes. *Phosphate buffered saline (PBS)* is a mixture of phosphate buffer (= monobasic potassium phosphate + dibasic potassium phosphate) and saline (= sodium chloride). DPBS is a modified version of PBS, where sodium phosphate and potassium chloride are also added.

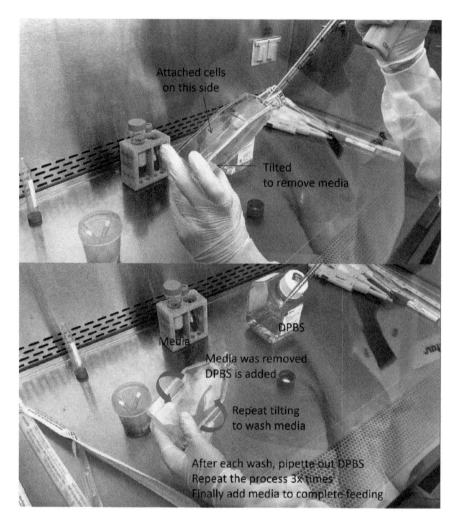

Fig. 3.9 Cell feeding procedure

14. Remove the old medium from the T-25 flask using a serological pipette. Tilt the T-25 flask to remove the old medium entirely. Use a waste beaker. Cells will remain on the surface of the T-25 flask (they are anchorage-dependent cells).
15. Tilt the flask to ensure the buffer solution is rinsing all areas of the flask properly.
16. Wash cells with 1 mL *DPBS*. This washing should get rid of any residual culture medium and waste products. Remove DPBS using a serological pipette. Tilt the T-25 flask to ensure buffer solution is rinsing all areas of the flask and entirely removing the DPBS. Repeat this step two times.
17. Slowly add warm media (described in steps 3–6) to the T-25 flask. Tilt the T-25 flask side-to-side to ensure the surface is adequately covered.
18. Culture for another day. Repeat cell imaging. Has the morphology and confluency been improved?

3.13   Laboratory Task 2: Cell Passaging

The overall procedure is illustrated in Figs. 3.10 and 3.11.

19. Repeat steps 14–16 in task 1 to remove culture medium and rinse twice with DPBS.
20. Place 1 mL of trypsin-EDTA solution into the T-25 flask. Incubate for 2–6 min at room temperature or in a CO_2 incubator at 37 °C (optional). Trypsin works better at 37 °C (body temperature) than at room temperature.
21. Add 1 mL of warm culture media (steps 3–6) to neutralize trypsin-EDTA.
22. Image the cells on the T-25 flask. Check morphology and counts of cells. Examples of bright-field microscopic images before and after trypsin-EDTA treatment are shown in Figs. 3.12 and 3.13.
23. Transfer 2 mL (1 mL of trypsin-EDTA solution and 1 mL of culture medium) from the T-25 flask into a 15-mL tube.
24. Centrifuge at 1,500 rpm (rounds per minute) for 5 min. Cells will precipitate out from the solution forming white cell pellets at the bottom.
25. While waiting for centrifuging, prepare 4 mL of media in a 37 °C water bath.
26. When centrifuging is finished, remove supernatants using a pipette. Avoid touching the cell pellets.
27. Resuspend the pellet with 1 mL warm medium.

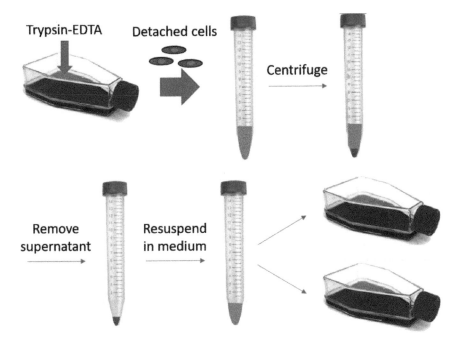

Fig. 3.10 Cell passaging process

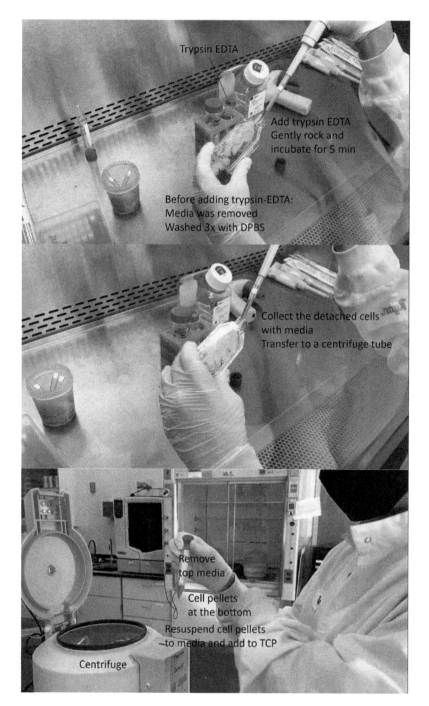

Fig. 3.11 Cell passaging procedure

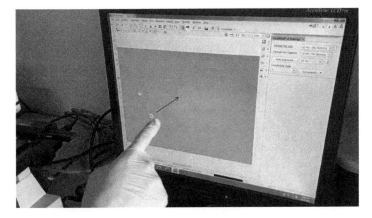

Fig. 3.12   Detached cell is freely moving on a T-25 flask

28. Take this cell suspension, and along with the remaining 3 mL warm medium, place them together in a new T-25 flask.
29. Gently shake the T-25 flask to allow the cells to spread over the flask.
30. Culture for another day. Repeat cell imaging.
31. Wrap-up: Clean biosafety cabinet with ethanol and paper towels. Turn the UV lamp on for 5–10 min.

Review Questions
1. What is the waste from glycolysis? What is the waste from glutaminolysis?
2. Is glycolysis aerobic or anaerobic? Is TCA cycle aerobic or anaerobic?
3. What is oxidative phosphorylation (respiration)?
4. What is deregulated metabolism?
5. What is GLUT1?
6. Discuss the impacts of low glucose feeding (< 1 mM) and low glutamine feeding (< 0.2 mM).
7. Discuss the impacts of high glucose feeding and high glutamine feeding.
8. When is the glutamine metabolism affected by glucose concentration? When is glycolysis affected by glutamine concentration?
9. When the oxygen level is low, what is the preferred substrate between glucose and glutamine?
10. Discuss the strengths and weaknesses of using a serum as a culture medium.
11. What is a defined medium, and what are the advantages?
12. What is cell feeding, and why do you need it?
13. What is cell passaging, and why do you need it?
14. How do you do cell passaging for anchorage-dependent cells and anchorage-independent cells?
15. Define specific growth rate (in an equation).

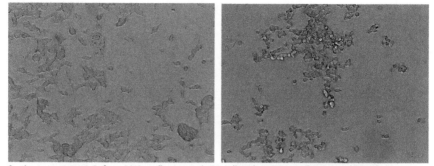

2-minute treatment. Before: 80% confluency and 122 cells. After: 70% confluency and 89 cells.

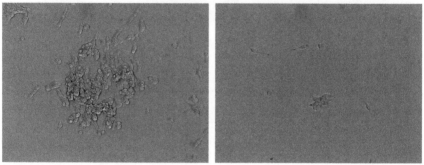

3-minute treatment. Before: 70% confluency and 115 cells. After: 20% confluency and 17 cells.

4-minute treatment. Before: 45% confluency and 102 cells. After: 15% confluency and 15 cells.

Fig. 3.13 Bright-field microscopic images (using a 10× objective) of anchorage-dependent mammalian cells on a T-25 flask before and after trypsin-EDTA treatments. Note the smaller number and rounded morphology of cells after trypsin-EDTA treatments.

Reference

Sigma-Aldrich. Product information: Dulbecco's Modified Eagle's Medium (DME). Catalog number D0422. https://www.sigmaaldrich.com/content/dam/sigma-aldrich/docs/Sigma/Formulation/d0422for.pdf. Accessed Jan 2020.

Chapter 4
Cell Imaging

In the previous chapters, we have learned the basics and details of cell culture. Once cell culture is complete, there is a need to image the cells. While simple light microscopy can be used, it is preferable to "stain" the specific components of cells and image them separately for most tissue engineering applications. Most mammalian cell cultures are imaged by such staining. Staining is typically done using both fluorescent dye and bioreceptor. This chapter will learn the details of cell imaging, especially the one based on fluorescent staining. Here are three inquiries for you:

Inquiry 1. What is fluorescence?

How are fluorescent dyes different from other colorimetric dyes?

Inquiry 2. What kind of bioreceptors can be used toward staining cells?

What subcellular components or proteins should be targeted and imaged?

Inquiry 3. How is fluorescence microscopy different from conventional bright-field microscopy?

What additional components are added to conduct fluorescence microscopy on a light microscope? How can we use a fluorescence microscope? Can we stain and image multiple subcellular components and proteins for the same cell culture sample?

4.1 Overview of Fluorescence Microscopy

For most mammalian cell cultures, including tissue engineering applications, it is preferable to image the cell culture using *fluorescence microscopy*. Subcellular components or proteins are stained using different fluorescent dyes. For example, nuclei can be stained with blue-fluorescent dye, actin filaments with green-fluorescent dye, and mitochondria with red-fluorescent dye. We have already

© Springer Nature Switzerland AG 2022
J.-Y. Yoon, *Tissue Engineering*, https://doi.org/10.1007/978-3-030-83696-2_4

Fig. 4.1 (Identical to Fig. 2.19) Images acquired by a fluorescence microscope. The blue fluorescence image shows nuclei (top left), the green fluorescence image shows actin filaments (top right), the red fluorescence image shows mitochondria (bottom left), and the stacked image is finally shown (bottom right). (Yoon, 2016. Reprinted with permission. (C) 2016 Springer)

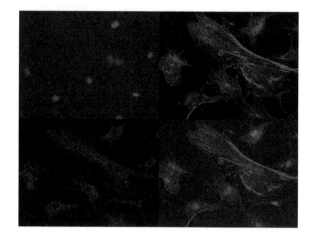

exercised such fluorescence microscopy in laboratory task 2 of Chap. 2 (Sect. 2.12). As it is typically challenging to image three *fluorescent dyes* altogether, fluorescent images are acquired one by one. For example, the blue-fluorescent image of nuclei is collected first, followed by the green-fluorescent image of actin filaments, and finally, the red-fluorescent image of mitochondria. These three images can be stacked together to generate a nice-looking picture, as shown in Fig. 2.20, although such a stacked picture is not real (Fig. 4.1).

Bioreceptor is necessary to bind specifically to a subcellular component or a protein. While many different bioreceptors can be used, antibodies are frequently used as such bioreceptors. *Antibodies* are proteins that can bind to specific antigens. *Antigens* are "foreign" substances that may cause problems in mammal or human bodies. Antigens can be proteins, polysaccharides, lipids, nucleic acids, viruses, bacteria, etc. Antibodies are produced by *B lymphocyte* (= *B cell*), which is one type of white blood cells (WBCs) and a part of the body's immune system. When an antibody is conjugated with a fluorescent dye and used to stain a specific protein or a subcellular component, the process is referred to as *immunostaining*, that is, staining process utilizing proteins from the immune system (= antibodies). Of course, non-antibody staining is also widely used, for example, *phalloidin* can bind specifically to actin filaments.

4.2   Fluorescence

Before we learn about fluorescent dyes, we need to understand fluorescence. Figure 4.2 illustrates the principle of fluorescence.

When atoms are irradiated with light, in this case blue light (a part of visible light, with wavelength ranging from 400 to 500 nm), their electrons absorb the light and move to a higher energy state (excited state). Usually, the excited electrons eventually lose the energy and return to the ground state. This energy can be

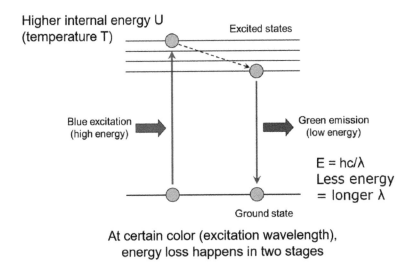

Fig. 4.2 Illustration of fluorescence. (Yoon, 2016. Reprinted with permission. (C) 2016 Springer)

converted to heat to increase its temperature, where the light is absorbed. Alternatively, it can simply be emitted in the same form of energy, that is, blue light, where the light is scattered.

However, for specific molecules, the excited electrons lose their energy a tiny bit for a short duration of time. When the electrons return to the ground state, they emit a smaller amount of energy. The energy of light (actually, all electromagnetic waves) is inversely proportional to the wavelength.

$$E = \frac{hc}{\lambda}$$

In this case, the emitted light is green color, whose wavelength is longer (500–550 nm) than that of blue color. This longer wavelength corresponds to the lower energy according to the above equation. Therefore, you can make a clear distinction between irradiated (or incident) light and emitted light, which is a significant benefit over colorimetric dyes.

In conventional *colorimetric dyes*, white light is irradiated to the sample. White light comprises all three primary colors (RGB): red from 550 to 750 nm, green from 500 to 550 nm, and blue from 400 to 500 nm. If you mix all three primary colors, it will look white. Colorimetric dyes "absorb" a specific range of wavelengths and convert it into a different energy form (mostly heat). If a colorimetric dye absorbs blue color but not green and red color, the resulting coloration will be green + red, that is, yellow color (under white light). Colorimetric dyes are typically quantified by measuring the *absorbance* at a specific wavelength. For the above colorimetric dye, blue absorption should be measured, although its coloration is yellow. To identify only the light from a colorimetric dye, you will need to measure the absorbance at one specific wavelength where its absorbance is maximum. If the cells, culture media, or protein products also absorb at the same wavelength, it will be challenging to identify and quantify only the colorimetric dye.

On the other hand, fluorescent dyes utilize two different wavelengths, one for *excitation* and the other for *emission*. For example, the well-known *fluorescein* is best excited at 494 nm (blue), and its emission is peaked at 521 nm (green). Therefore, it can make a clear distinction between the fluorescent dye and all other molecules in the sample. The chance for the cells, culture media, or proteins to exhibit the same fluorescence is relatively low. The fluorescence phenomenon can be found in certain specific types of molecules, which will be discussed in the following section.

The difference between the excitation and emission wavelengths is referred to as *Stokes shift*. When a more considerable extent of energy is lost for a short duration, the Stokes shift will be greater, which is preferred for distinguishing between excitation and emission light.

Figure 4.3 shows the example spectra of a fluorescent dye solution – in this case, fluorescein. Under ambient lighting, that is, white light, it absorbs blue light and exhibits green + red = yellow coloration. When the solution is irradiated with blue light, some blue light is used (= absorbed) to excite fluorescein molecules. Fluorescein eventually emits green light (= green fluorescence). When the detector is aligned in parallel with the light source, it picks up both the excitation (blue) and emission (green), where the emission is typically weaker than the excitation. When the detector is not aligned to the light source or a pair of optical bandpass filters are used (which allow only a specific narrow range of wavelengths to pass through, typically ±5 or ±10 nm range), the detector picks up only the green fluorescence

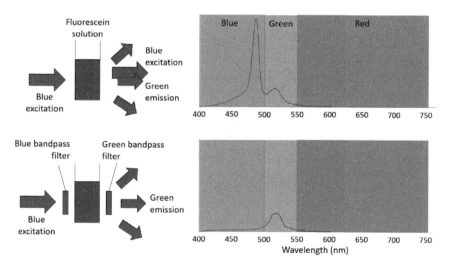

Fig. 4.3 Fluorescence spectra of a fluorescein solution

emission. In fluorescence-based cell imaging, optical bandpass filters are always used, which will be discussed in the later section.

4.3 Fluorescent Dyes

Many different fluorescent dyes are available for biosensing as well as cell imaging. The three most well-known fluorescent dyes for cell imaging would be DAPI, FITC, and TRITC.

DAPI is short for 4′,6-diamidino-2-phenylindole (Fig. 4.4). Its excitation peak is 358 nm, which is in the *ultraviolet* (*UV*) region of electromagnetic waves, shorter than blue light. Its emission peak is 461 nm, which is in the blue part. It is typically referred to as a blue-fluorescent dye considering its emission color. While DAPI itself is a fluorescent dye, it also functions as a bioreceptor. It binds to DNA and thus stains nuclei of cells.

Specifically, DAPI binds to A-T-rich regions of DNA. [There are four different bases in DNA: adenine (A) and guanine (G) are the larger purines, while thymine (T) and cytosine (C) are the smaller pyrimidines. A always binds to T, while G binds to C.] For many other fluorescent dyes, they need to be conjugated with a bioreceptor to stain specific subcellular components or proteins. DAPI does not require a separate bioreceptor as DAPI itself is also a bioreceptor. As nuclei staining is essential in identifying their locations and counting the number of cells, DAPI staining is very popularly used in cell imaging. Other dyes like DAPI are also available (as both fluorescent dye and bioreceptor).

FITC is short for *fluorescein isothiocyanate*. FITC is a variant of *fluorescein*, where isothiocyanate (–N=C=S) is added to the lower aromatic ring's bottom. The structures of fluorescein and FTIC are shown in Fig. 4.5. As mentioned in the above paragraph, most fluorescent dyes are conjugated with a bioreceptor to stain a subcellular component or a protein specifically. Unfortunately, fluorescein itself is not very soluble in water and not very reactive with bioreceptors. Fluorescein is typically modified with the highly reactive isothiocyanate group (there are two highly reactive double bonds in isothiocyanate).

Fluorescein's (and FITC's) excitation peak is 483 nm, which is in the blue part, and its emission peak is 520 nm, which is in the green part. Therefore, it is a green-fluorescent dye. It does not function as a bioreceptor – note that DAPI is an

Fig. 4.4 DAPI

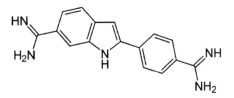

Fig. 4.5 Fluorescein (left) and FITC (right)

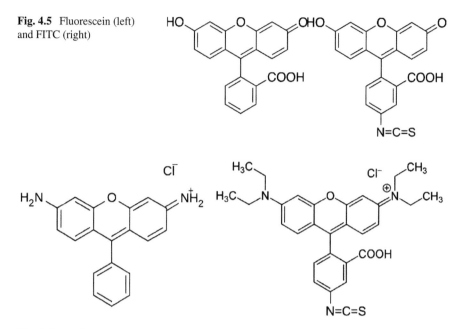

Fig. 4.6 Rhodamine (left) and TRITC (right)

exception – and requires the conjugation with a bioreceptor (thus the need for FITC). Also, note the overall large size of the molecular structure of fluorescein (and FITC) over DAPI and the higher number of resonating aromatic ring structures of fluorescein (and FITC) over DAPI. And the excitation and emission wavelengths are longer with fluorescein (and FITC) than with DAPI.

TRITC is short for *tetramethylrhodamine isothiocyanate*, which is a modified form of rhodamine. The structures of rhodamine and TRITC are shown in Fig. 4.6. Like FITC, isothiocyanate (–N=C=S) is added to the bottom of the lower aromatic ring of rhodamine. It facilitates its conjugation to a bioreceptor. Rhodamine's (and TRITC's) excitation peak is 543 nm, which is in the green part, and its emission peak is 580 nm, which is in the red region, thus a red-fluorescent dye. As 580 nm is quite close to green color and the human eye's green corneal cells can also recognize this wavelength to a certain extent, it will look yellow-red to a human eye (and most digital cameras). Also, note the overall size of rhodamine (and TRITC) is even bigger than fluorescein (and FITC), correlating to the longer excitation and emission wavelengths.

Another popular variant of rhodamine dye is *tetramethylrhodamine methyl ester (TMRM)*. Its structure is shown in Fig. 4.7. As it is a variant of rhodamine, it is still excited with green color (548 nm) and emits red color (574 nm), like those of rhodamine and TRITC. It is subsequently a red-fluorescent dye. The differences of TMRM from TRITC are that the –COOH in the lower aromatic ring is substituted with –COOCH$_3$, thus no longer acidic, and that it is prepared as a salt with perchlorate (ClO$_4^-$) rather than with chloride ion (Cl$^-$).

Fig. 4.7 TMRM

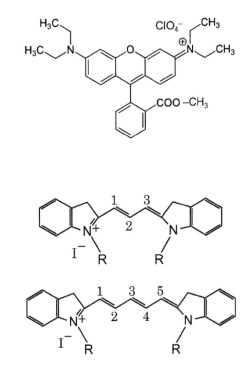

Fig. 4.8 Cy3 (top) and Cy5 (bottom). (Yoon, 2016. Reprinted with permission. (C) 2016 Springer)

Mitochondria inside a cell, which we have learned in Chap. 3, are essential in cell metabolism (especially the TCA cycle's oxidative phosphorylation) and often indicate cells' healthiness. While mitochondria are subcellular components (*organelle*), they still have membranes with electrical potential differences across their membranes (if they are healthy). TMRM can permeate through the mitochondrial membrane toward its interior utilizing this potential difference, thus effectively staining mitochondria. Like DAPI, TMRM functions as both a fluorescent dye and a bioreceptor.

While most fluorescence microscopes are designed to work with these three basic fluorescent dyes, that is, DAPI, FITC, and TRITC, many other fluorescent dyes are also available. A good example is cyanine dyes, which bear similar structures with varying molecular sizes (varying excitation and emission wavelengths). Figure 4.8 shows two examples, Cy3 and Cy5. *Cy3* has an excitation peak at 550 nm (the boundary of green and red regions) and an emission peak at 649 nm (red). *Cy5* has an excitation peak at 570 nm (yellow-red; refer to the description of TRITC) and an emission peak at 670 nm (red).

Question 4.1 When the chemical structures are provided, can you sort multiple fluorescent dyes in ascending order of their emission wavelengths?

4.4 Bioreceptors

We have already learned that some fluorescent dyes can also be used as bioreceptors. DAPI, a blue-fluorescent dye, can bind to DNAs (specifically the A-T-rich regions) and thus staining cell nuclei (Fig. 4.9). TMRM, a red-fluorescent dye, can bind to mitochondrial membranes and be accumulated inside mitochondria, thus staining mitochondria within cells (Fig. 4.10). Refer to Fig. 4.1 for the sample microscopic images of DAPI and TMRM staining.

However, for most fluorescent dyes, a specific bioreceptor is necessary to stain specific subcellular components or proteins inside cells. While most receptors used in cell imaging are antibodies, non-antibody bioreceptors are more commonly used. The most frequently used example is *phalloidin*, which can specifically bind to actin filaments (Fig. 4.11). With this staining, you can check the morphology and distribution of actin filaments and their focal adhesion links. Focal adhesion has already been explained briefly in Chap. 2, and details will be discussed in Chap. 7. Phalloidin can also be conjugated to FITC or TRITC. You can find a microscopic sample image of phalloidin-FITC staining in Fig. 4.1.

Fig. 4.9 DAPI's binding to A-T-rich regions of DNA

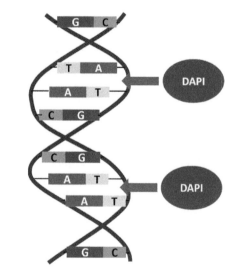

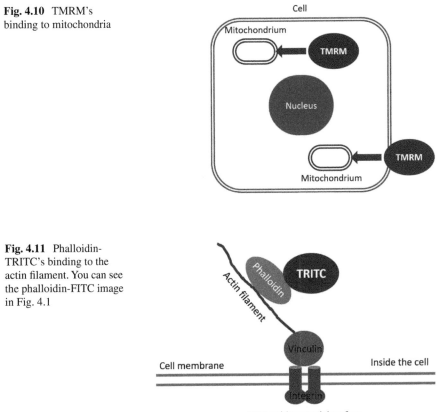

Fig. 4.10 TMRM's binding to mitochondria

Fig. 4.11 Phalloidin-TRITC's binding to the actin filament. You can see the phalloidin-FITC image in Fig. 4.1

Numerous antibodies can be used for cell imaging. For example, an antibody to tubulin (*anti-tubulin*) can be conjugated to FITC or TRITC to stain microtubules (refer to Chap. 2). Similarly, an antibody to vinculin (*anti-vinculin*) can be conjugated to FITC or TRITC to stain vinculin, which can only be found at the point of focal adhesion (refer to Chap. 2). As vinculins are found where integrins and actin filaments are connected, they can often be found at the same locations. When imaging vinculin and actin filaments together, different fluorescent dyes should be used, for example, anti-vinculin-FITC and phalloidin-TRITC, or anti-vinculin-TRITC and phalloidin-FITC.

You may notice that most bioreceptors are conjugated with either green-fluorescent dye (e.g., FITC) or red-fluorescent dye (e.g., TRITC), but not with blue-fluorescent dye. The reason is that DAPI staining is almost ubiquitously used to identify the location of cell nuclei or to count the number of cells, leaving only two other colors (green and red) out of three primary colors (RGB).

Many other bioreceptors can be used for staining various proteins or subcellular components within cells. Many bioreceptors are commercially available from multiple vendors, and it is unnecessary to list them all. A small number of commonly used bioreceptors, for example, phalloidin, are available with fluorescent dye preconjugated by commercial vendors, for example, phalloidin-FITC or phalloidin-TRITC. However, most other bioreceptors, especially antibodies, are not available with fluorescent dye pre-conjugated. It is not practical to conjugate fluorescent dyes to an extensive array of bioreceptors (mostly antibodies). A more realistic approach is using "antibody to antibody" or *secondary antibody* pre-conjugated with a fluorescent dye. For example, if anti-vinculin (*immunoglobulin G or IgG*; the most common form of antibodies) is made from mouse cells, we can use rabbit antibody against all mouse antibodies, for example, rabbit anti-IgG, that is pre-conjugated with FITC. As the number of mammal species used for producing antibodies is relatively small, for example, mouse, rat, goat, rabbit, horse, human, etc., commercial vendors need to conjugate fluorescent dyes to a small number of anti-IgGs. Figure 4.12 graphically illustrates the use of a secondary antibody for vinculin staining.

Question 4.2 Name all proteins (bioreceptors and secondary antibodies, if necessary) and all fluorescent dyes used for staining:

A. Nucleus
B. Actin filament
C. Microtubule
D. Mitochondria
E. Vinculin

Question 4.3 To image the following combinations of subcellular components in a single stacked image, choose the appropriate proteins and fluorescent dyes for each element:

A. Nucleus, actin filament, and vinculin
B. Nucleus, actin filament, and microtubule
C. Nucleus, actin filament, and mitochondria

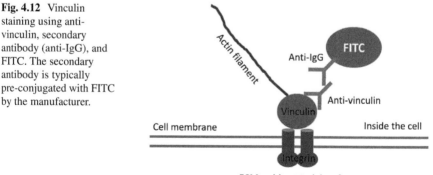

Fig. 4.12 Vinculin staining using anti-vinculin, secondary antibody (anti-IgG), and FITC. The secondary antibody is typically pre-conjugated with FITC by the manufacturer.

4.5 Fluorescence Microscope

Once the fluorescent dye and bioreceptor (if necessary) are prepared, which usually comes in as a kit, you will need to know how to use a fluorescence microscope. Figure 4.13 shows an inverted microscope photograph, where the objective is located underneath the sample stage. If the objectives are located on top of the sample stage, it is called an upright microscope. As the inverted microscope provides more room for a user to work with the sample, it is more prevalent in cell imaging. The *objective* is a crucial element of a microscope that magnifies the images, consisting of lenses or lenses with other optical components. Most microscopes can accommodate multiple objectives (10×, 20×, 40×, 60×, 100×, etc.), where they can be easily rotated to select the desired objective.

A *fluorescence microscope* is a one type of light microscope, where the following components are added to it: (1) an excitation light source, (2) an excitation optical bandpass filter (or *excitation filter*), (3) a dichroic mirror, (4) an emission optical bandpass filter (or *emission filter*), and (5) software application to process fluorescence images. Most benchtop light microscopes are designed to accommodate the above components, and you can convert them into fluorescence microscopes. Figure 4.14 shows the diagram of these added components for an inverted fluorescence microscope.

Let me explain the working principle of a fluorescence microscope for imaging phalloidin-FITC, where excitation should be blue and emission should be green. A white light source can be used as an excitation source, as the excitation filter can allow only the desired color or the range of wavelength. An incandescent light bulb is not the right choice, as it cannot produce UV that is necessary for DAPI-based nuclei staining. The same goes for a halogen lamp. The popularly used "white" excitation sources are either a *xenon arc lamp* or a *mercury-vapor lamp*. Both produce UV as well as most visible light. *LED* and *laser* are also popularly used, where they generate only a specific color. A laser is particularly advantageous as its

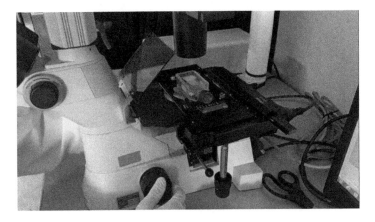

Fig. 4.13 An inverted microscope

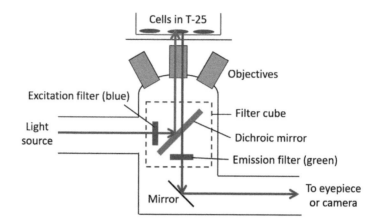

Fig. 4.14 A schematic illustration of an inverted fluorescence microscope

spectrum is exceptionally narrow, developing monochromatic light and eliminating an excitation filter's need.

Regardless of the excitation source, a blue light will hit the *dichroic mirror*. For a FITC-specific dichroic mirror, it should reflect any light shorter than 500 nm (i.e., blue color) and pass light that are longer than 500 nm (i.e., green color). At this time, blue light should be reflected, travel through an objective, and hit the cell culture sample. FITC is properly excited and emits green fluorescence in all directions. A small portion of this green fluorescence should return to the objective (with proper magnification) and hit the dichroic mirror. While blue excitation light tends to travel toward the upward direction, a small portion of it may return to the dichroic mirror. As described above, a dichroic mirror should allow green fluorescence to pass through. The emission filter at the bottom chooses only the green fluorescence emission and eliminates any blue excitation light.

The excitation filter, dichroic mirror, and emission filter come as a set. This set is called a *filter cube*. There is a filter cube for FITC, optimized for its peak excitation and peak emission wavelengths. You can have another filter cube for DAPI and the third one for TRITC. When you buy a fluorescence microscope, these three filter cubes come as a default option. These three filter cubes can be loaded altogether into a single fluorescence microscope, and you can slide them through to choose one. Because of this, three slots are typically available. Cheaper microscopes come with only two slots, where you must manually load and unload filter cubes if you want to image three fluorescent dyes. For nontraditional dyes, for example, Cy3 and Cy5, you will need to purchase a new set of filter cubes designed explicitly for them.

Images can be confirmed in two different ways. There is an eyepiece that you can see the fluorescent images with your naked eye. You can also connect a stand-alone digital camera or a digital camera system ("detector") connected to a separate computer. A digital camera is more versatile as it can adjust the brightness, contrast, and, most importantly, exposure time. When the fluorescent emissions are very weak, you will not be able to "see" them through the eyepiece. However, with the digital camera, you can extend the exposure time for a long time to obtain much brighter and clearer fluorescence images.

With a fluorescence microscope, you can do a wide variety of analyses for cell culture. The following list is just a few examples and not exhaustive.

(1) Cell counts in the *field of view* (*FOV*). Cells can typically be counted with DAPI staining through counting the number of nuclei. It may be possible to use a specific kit to make a distinction between live and dead cells.
(2) Confluency. Surface coverage of cells can be evaluated with DAPI staining (to count the cells) and/or phalloidin staining (to identify the cytoskeleton).
(3) Cell metabolism. Various enzymes, proteins, wastes, etc., can be stained and imaged. One of the easiest ways is TMRM staining that can tell you the level of mitochondrial activity.
(4) Protein expression. When the cells are designed to produce a specific protein product, it can be stained to identify and quantify its successful production.
(5) Cell morphology. The shape and structure of cells can be evaluated, for example, with phalloidin staining to identify actin filaments.
(6) Focal adhesion. Vinculin can be stained to identify the locations and extent of focal adhesion.

4.6   Photobleaching

Fluorescent dyes gradually lose their ability of fluorescence upon repeated exposure to excitation light. As a result, fluorescence images become "dim," as shown in Fig. 4.15. This phenomenon is called *photobleaching*, which came from bleach that can remove the color from fabric or stain from clothes. The best way to reduce photobleaching is to minimize the exposure time to the excitation light, that is, not to turn on the excitation light when the sample is not being imaged. Other ways are (1) shortening of exposure time, (2) decreasing the intensity of excitation, and (3) using a fluorescent dye that is less prone to photobleaching.

Quantum dots have been suggested as a solution to this photobleaching problem. They are semiconductor nanoparticles that exhibit *artificial fluorescence*. Excitation is almost exclusively UV, while emission can be varied from blue, green, red, and even near-infrared (NIR) based on their sizes, which significantly simplifies the imaging of two or more stains. Most importantly, photobleaching is not a significant issue with quantum dots. Hence, quantum dots seem to be a superior choice over fluorescent dyes, although they have their issues. The biggest problems of quantum dots in cell imaging are that (1) they are more unstable than fluorescent dyes and (2) they cannot easily be conjugated to bioreceptors. At the time of writing, fluorescent dyes are still being widely used in cell imaging, while quantum dot use is somewhat limited.

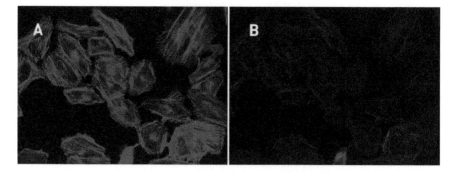

Fig. 4.15 Normal (left) and photobleached (right) fluorescence microscopic images

4.7   Smartphone Fluorescence Microscope

These days, smartphones are equipped with high-quality digital cameras, which have eradicated most low- and medium-end digital camera markets. Most digital cameras have a *zoom* feature, where the user can magnify images. *Digital zoom* is not an actual magnification as it merely enlarges the image at the cost of sacrificing image resolution. On the other hand, *optical zoom* is an actual magnification by adjusting the distance or position of lenses. This mechanism is quite like that of an *objective*, which is the crucial component of all light microscopes. Optical zoom had not been available in smartphones in the past. It has become a standard feature of recent smartphones. However, its magnification is still quite limited (around 2×–4× at the time of writing). The use of dual or triple cameras can provide an additional zoom effect, although they are not equivalent to the true optical zoom. Considering that most recent smartphones are relatively thin, it will be challenging to implement a true optical zoom comparable to a benchtop light microscope.

At the time of writing, it becomes necessary to attach additional objectives to use a smartphone as a microscope. They are referred to as *microscope attachment to smartphone* or *microscope phone adaptor*, or *smartphone microscope*. Figs. 4.16 and 4.17 show a couple of examples. As newer smartphone models are announced every year, smartphone microscopes are also being modified on an annual basis. By the time this book is published, the following examples may not be available in the market.

As fluorescence microscopes are modified versions of light microscopes, it is possible to convert these smartphone microscopes into *smartphone fluorescence microscopes*. Unlike benchtop light microscopes, smartphone microscopes are not designed to accommodate filter cubes. Therefore, bandpass and dichroic filters must be installed manually. One example is shown in Fig. 4.18 on the left, where an LED is used as an excitation light source. The T-shaped plastic housing (3D-printed) includes two lenses, two bandpass filters (one near the LED as an excitation filter and the other near the smartphone camera as an emission filter), and a dichroic filter at the T-junction. It has

Fig. 4.16 Various approaches of smartphone microscopes, using a single ball lens (**a**), multielement lens (**b**), and an objective lens (**c**). (Zhu et al., 2021. Reprinted with permission, (C) 2021 Elsevier)

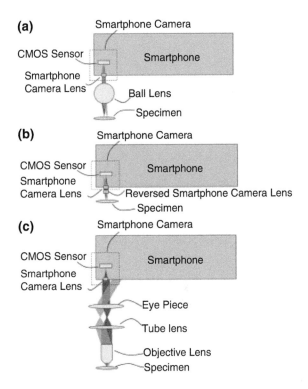

Fig. 4.17 Smartphone microscope from 4D Optical LLC. (Reprinted with permission from 4D Optical LLC, http://microbescope.com/, accessed January 2021)

three components – objective, filter cube, and light source – built into a single attachment. Another example is shown in Fig. 4.18 on the right. A bandpass emission filter is attached (taped) to the commercial microscope attachment, while an LED is irradiated from the side as excitation light. As it is being illuminated from the side, most excitation light will not be picked up by the objective.

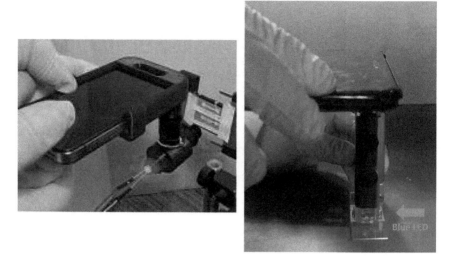

Fig. 4.18 Examples of smartphone fluorescence microscopes. Left: a 3D-printed custom-built attachment incorporating an objective (two lenses), a filter cube (two bandpass filters and a dichroic filter), and an LED excitation source. (Fronczek et al., 2014. (C) Open access article distributed under the terms of the Creative Commons Attribution 4.0 International License). Right: a bandpass emission filter is taped at the end of a commercial microscope attachment, while a separate blue LED is irradiated from the side. (Chung et al., 2019. (C) Open access article distributed under the terms of the Creative Commons Attribution 4.0 International License)

4.8 Laboratory Task 1: Fluorescence Imaging of Nucleus and Cytoskeleton

Imaging cells' nuclei with DAPI and actin filaments (the most crucial cytoskeleton) with phalloidin-FITC or phalloidin-TRITC are the two most performed fluorescence imaging methods for mammalian cell culture. DAPI staining can count the number of cells and subsequently evaluate confluency (together with phalloidin staining). Phalloidin staining can be used to identify cells' morphology and confluency (together with DAPI staining).

In this task, mammalian cells should be cultured before the experiments. Repeat Objective 1 of Chap. 3, but this time on a *microwell plate* or a *well plate*. It is a flat plate with multiple wells on it, typically with 96 wells (12 × 8) (Fig. 4.19). Each well serves as a small flask. As many as 96 different experiments or reactions can be performed simultaneously. Microwell plate is useful for conducting multiple experiments under an identical environment. It is handy for numerous mammalian cell cultures for 10–20 groups (e.g., in a teaching laboratory) while using only a single CO_2 incubator.

Objective 1. Staining with DAPI and Phalloidin-TRITC
 1. At the end of your desired incubation period, while the cells are still enclosed in the CO_2 incubator, sterilize the closed biosafety cabinet with UV for 30 min,

Fig. 4.19 A microwell plate

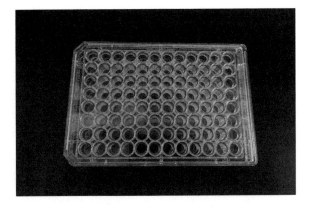

then open the front window and wait 10 min to allow the laminar airflow to stabilize. Once ready, put on gloves and spray your hands and forearms with ethanol. Properly spray materials with ethanol before entering the biosafety cabinet.

2. Remove the microwell plate from the CO_2 incubator, spray with ethanol, and transfer to the biosafety cabinet. Be careful not to spray directly at the edges where ethanol may be introduced into the microwell plate.

3. Using a pipette, remove and discard all waste media from each well (containing cells) and *immediately* add 100 μL of 4% paraformaldehyde. This addition kills and fixes the cells to the surface. Allow sitting at room temperature in the biosafety cabinet for 15 min (Fig. 4.20).

4. Remove all paraformaldehyde and rinse each well twice (100 μL per rinse) with washing buffer.

5. Perforate the cells' membranes by adding 100 μL of 0.1% Triton X-100 (surfactant) to each well for 5 min at room temperature (Fig. 4.21).

6. Remove all Triton X-100 and rinse each well twice with washing buffer.

7. Apply 100 μL blocking buffer to each well. Allow incubating at room temperature in the biosafety cabinet for 30 min (Fig. 4.22).

8. In a 2 mL centrifuge tube, dilute the TRITC-conjugated phalloidin (phalloidin-TRITC) in 1× PBS to an approximate dilution of 1:250 (for 250 μL, add 1 μL TRITC to 249 μL blocking buffer). Add 100 μL of this solution to each well and incubate at room temperature for 30 min (Fig. 4.23).

9. Remove the phalloidin-TRITC solution from each well and wash three times with washing buffer, with 5 min of incubation at room temperature for each wash cycle.

10. In a 2 mL centrifuge tube, dilute the DAPI nuclei counterstain in 1× PBS to an approximate dilution of 1:250 (for 250 μL, add 1 μL DAPI to 249 μL blocking buffer). Add 100 μL of DAPI solution to each well and incubate at room temperature for 5 min (Fig. 4.24).

11. Remove DAPI solution from each well and add 100 μL 1× PBS to prevent the treated cells from drying. Seal the edges of the microwell plate with parafilm

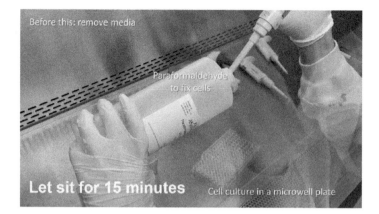

Fig. 4.20   Adding paraformaldehyde to fix cells in a microwell plate

Fig. 4.21   Adding Triton X-100 to perforate cell membranes

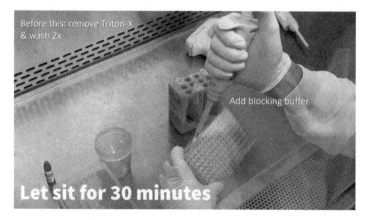

Fig. 4.22   Adding blocking buffer

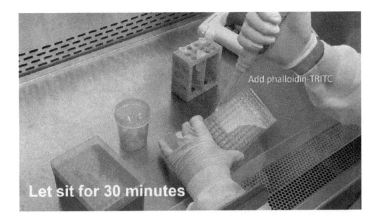

Fig. 4.23 Adding phalloidin-TRITC

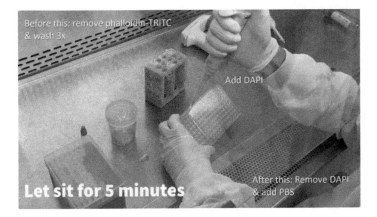

Fig. 4.24 Adding DAPI

(sealing film). Keep the microwell plate covered in aluminum foil and place it in the refrigerator if not imaging immediately.

Objective 2. Fluorescence Microscopy

12. In the darkroom, power on the microscope, the fluorescent light source, the camera attachment, and the attached computer with imaging software. Ensure that the fluorescent light aperture is closed or that the light path is blocked. This blockage will help to prevent photobleaching of the fluorescent dyes until we are ready to view them.

13. Select the microscope's objective and viewing lens appropriate for viewing your cells and position it under the viewing stage. A 40× objective with a 10× viewing lens = 400× magnification is the right choice for a broad view of the cells. Position the well in a microwell plate directly over the microscope's objective (Fig. 4.25).

Fig. 4.25 Rotate to choose the desired objective. Place the selected individual well on top of the objective. Use the stage controller to move the stage in the x- and y-axis directions to locate the cells within each well

14. Turn on the light source and switch the viewing path to the microscope's viewing lens (*eyepiece* or *binocular*) (Fig. 4.26). Through the viewing lens, you should be able to see the cells. Bring the cells into focus (Fig. 4.27). Once they are focused, ensure that the darkroom is sealed and turn off the light source and other lights in the room.

15. *Imaging the cell's nuclei with DAPI*: Position the DAPI filter cube into the fluorescent light source path and open the aperture, allowing the DAPI excitation light (358 nm) to reach the well plate. Through the eyepiece, blue oval light (461 nm) should emit throughout a dark background, which indicates the presence of cells. Each blue oval represents a single nucleus of a cell. *Safety note*: Be careful not to look directly at the light being emitted by the light source when the DAPI filter cube is in place as it is excited with UV light (358 nm) that could lead to eye damage.

16. Close the fluorescent light aperture, again blocking the light to prevent photobleaching of the DAPI stain. Switch the viewing path of the microscope to the camera and turn on the computer imaging software.

17. When you are prepared to take a photo, open the fluorescence light source aperture again and adjust the view's brightness by adjusting the camera software gain and exposure time. A brighter picture can be obtained by increasing the gain and prolonging the exposure time. Note that prolonged exposure to fluorescence excitation would lead to photobleaching. Capture and save your desired image and close the fluorescent aperture.

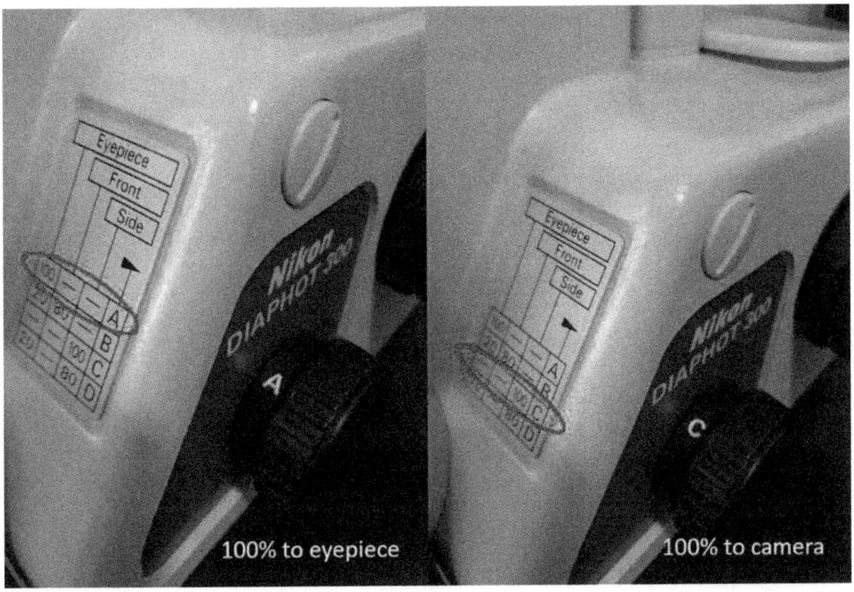

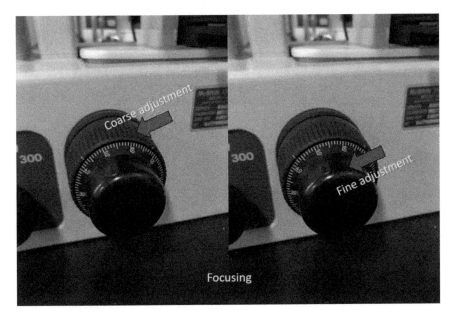

Fig. 4.26   Switching the viewing path to 100% eyepiece or 100% camera

Fig. 4.27   Coarse and fine focusing

18. *Do not move* the well plate after capturing your first image (DAPI). Phalloidin-TRITC stained photos will be captured exactly at the same location so that you can overlay these two images together.

19. *Imaging actin filaments with phalloidin-TRITC:* Replace the FITC filter with the TRITC filter and change the microscope's viewing path back to the microscope viewing lens (eyepiece or binocular).

20. Open the aperture, allowing the TRITC excitation light (543 nm) to reach the well plate. Through the eyepiece, spindly networks of red lines (580 nm) should radiate out in clusters throughout the dark background, which indicates the actin filaments within the cells.

21. Close the fluorescent light aperture, again blocking the light to prevent photobleaching of the TRITC stain. Switch the viewing path of the microscope to the camera and turn on the computer imaging software.

22. When you are prepared to take a photo, open the fluorescent light source aperture again and readjust the view's brightness by adjusting the camera software gain and exposure time.

23. Capture and save your second image and close the fluorescent aperture.

Objective 3. Stacking Fluorescent Images

24. Download and open ImageJ, a free image analysis software package from the U.S. National Institutes of Health. You can find it at the following site: http://imagej.nih.gov/ij/

25. Open your two fluorescent image files at once in ImageJ.

26. To merge the three images, go to the toolbar and follow the pathway:
    Image > Color > Merge Channels

27. In the red (C1), green (C2), and blue (C3) channels, select your TRITC (red) and DAPI (blue) image files. Check the boxes for Create Composite and Keep Source Images, and then select OK (Fig. 4.28). The result is an RGB image displaying the stained cells with a distinct nucleus and actin filaments.

Representative images are shown in Fig. 4.29. Note that raw DAPI and phalloidin-TRITC images are in gray scale. RGB color images are split into three color channels, each in 8-bit and a total of 24-bit. As fluorescence emission occurs in one color, only one channel is utilized while the other two not being used. Grayscale images can utilize all available channels, for example, 16- or 24-bit, thus providing better resolution and dynamic range. As we already know the color of fluorescence emission, grayscale images are typically sufficient. When stacking (merging) these two images, pseudo-colors (blue for DAPI and red for phalloidin-TRITC) are assigned for visualization convenience. However, these pseudo-colors are not strictly accurate colorations.

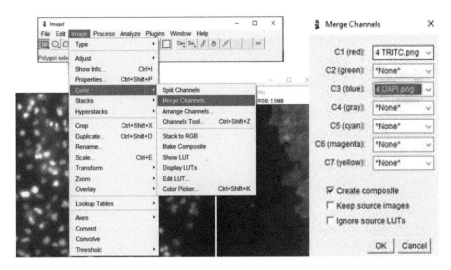

Fig. 4.28 Stacking multiple fluorescent images using ImageJ

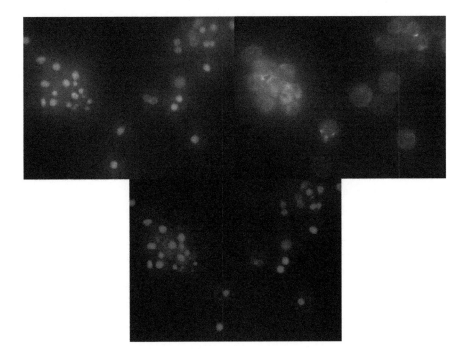

Fig. 4.29 Fluorescent images of DAPI- (top left), phalloidin-TRITC (top right), and stacked (bottom) fluorescent images of mammalian cells

4.9 Laboratory Task 2 (Alternative): Fluorescence Imaging of Nucleus, Cytoskeleton, and Focal Adhesion

This laboratory task is an alternative, advanced version of laboratory task 1, where vinculin staining is added. Vinculin can be found only where the focal adhesions exist. Anti-vinculin will be added first and rinsed. Secondary antibody-FITC will then be added together with phalloidin-TRITC.

After step 7, do the following:

28. In a 2 mL centrifuge tube, dilute the anti-vinculin solution in a blocking buffer to a dilution of 1:500 (for 500 μL, add 1 μL anti-vinculin to 499 μL blocking buffer). Add 100 μL of the anti-vinculin solution to each well and incubate at room temperature for 1 h. Remove the anti-vinculin solution from each well and wash three times with washing buffer, with 5 min of incubation at room temperature for each wash cycle.

Steps 8 and 9 should be substituted with the followings:

29. In a 2 mL centrifuge tube, dilute the FITC-conjugated secondary antibody (secondary antibody-FITC) and the TRITC-conjugated phalloidin (phalloidin-TRITC) in 1× PBS to an approximate dilution of 1:250 (for 250 μL, add 1 μL FITC and 1 μL TRITC to 248 μL blocking buffer). Add 100 μL of combined dye solution to each well and incubate at room temperature for 30 min.
30. Remove combined dye solution from each well and wash three times with washing buffer, with 5 min of incubation at room temperature for each wash cycle.

Add the following steps to Objective 2:

31. Imaging vinculin with anti-vinculin and secondary antibody-FITC: Replace the TRITC filter with the FITC filter and change the microscope's viewing path back to the microscope viewing lens (eyepiece or binocular).
32. Open the aperture, allowing the FITC excitation light (483 nm) to reach the well plate. Through the eyepiece, small points of green light (520 nm) should emit throughout a dark background, indicating the cells' focal adhesion sites.
33. Close the fluorescent light aperture, again blocking the light to prevent photobleaching of the FITC stain. Switch the viewing path of the microscope to the camera and turn on the computer imaging software.
34. When you are prepared to take a photo, open the fluorescent light source aperture again and readjust the view's brightness by adjusting the camera software gain and exposure time.
35. Capture and save your third image and close the fluorescent aperture.

In step 27, add FITC (green) image files as well as TRITC (red) and DAPI (blue) files.

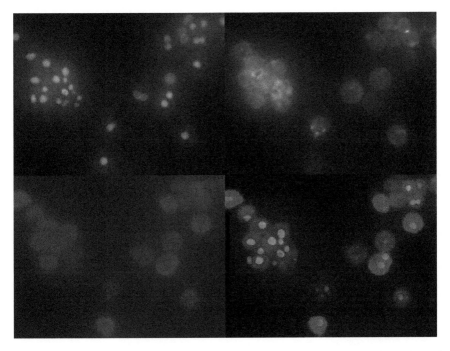

Fig. 4.30 Fluorescent images of DAPI- (top left), phalloidin-TRITC (top right), anti-vinculin and secondary antibody-FITC stained (bottom left), and stacked (bottom right) fluorescent images of mammalian cells. DAPI- and phalloidin-TRITC-stained photos are identical to those shown in Fig. 4.29

Additional raw FITC (vinculin) image and the stacked images with all three stains are shown in Fig. 4.30. Note that vinculins (green) are found only where actin filaments (red) are located, clearly indicating vinculin's binding to actin filaments.

Review Questions
1. Describe the differences between fluorescent dyes and colorimetric dyes.
2. Describe how a fluorescence microscope works.
3. Why do you need optical bandpass filters in fluorescence microscopy?
4. What is a filter cube?
5. Describe how to stain cell nuclei.
6. Describe how to stain focal adhesion sites.
7. Describe how to stain cytoskeleton (actin filaments and microtubules).
8. Describe how to stain mitochondria.
9. What protein is used in identifying focal adhesion?
10. Why do you need a secondary antibody for antibody staining?
11. What is photobleaching? How can you prevent it?
12. Why are fluorescence microscope images initially in gray scale?

References

Chung, S., Breshears, L. E., Perea, S., Morrison, C. M., Betancourt, W. Q., Reynolds, K. A., & Yoon, J. Y. (2019). Smartphone-based paper microfluidic particulometry of norovirus from environmental water samples at single copy level. *ACS Omega, 4*, 11180–11188. https://doi.org/10.1021/acsomega.9b00772

Fronczek, C. F., Park, T. S., Harshman, D. K., Nicolini, A. M., & Yoon, J. Y. (2014). Paper microfluidic extraction and direct smartphone-based identification of pathogenic nucleic acid from field and clinical samples. *RSC Advances, 4*, 11103. https://doi.org/10.1039/c3ra47688j

Yoon, J.-Y. (2016). *Introduction to biosensors* (2nd ed., Chapter 9). https://doi.org/10.1007/978-3-319-27413-3_9

Zhu, W., Gong, C., Kulkarni, N., Nguyen, C. D., & Kang, D. (2021). Chapter 9 – Smartphone-based microscopes. In J. Y. Yoon (Ed.), *Smartphone based medical diagnostics*. Elsevier. https://doi.org/10.1016/B978-0-12-817044-1.00009-0

Chapter 5
Stem Cells

In the previous chapters, we have learned the basics of mammalian cell culture, how to culture them (metabolism, feeding, and passaging), and how to image them (fluorescence microscopic imaging). For tissue engineering applications, special types of mammalian cells are often used, which can differentiate into different cell types. They are called stem cells. Here are some preliminary inquiries for you.

Inquiry 1. Have you heard about stem cells? If so, can you explain the differences between stem cells and normal cells?

Inquiry 2. How can you collect or obtain stem cells?

Can you "harvest" them from a specific organ or tissue? Can you "engineer" cells to produce stem cells?

Inquiry 3. Have you heard about ethical issues in harvesting and using stem cells?

Can you find out the regulations and laws of harvesting and using stem cells in your state or country? What about the other major countries?

5.1 What Are Stem Cells?

Stem cells are undifferentiated and unspecialized cells. They can be induced to become mature cells with a specific function. Mature cells can be used as an antonym to stem cells, but a more appropriate antonym should be *somatic cells*. Adult cells are often confused with somatic cells. While most adult cells are somatic cells, some do not, as there are adult stem cells.

Stem cells are essential not only in embryonic development but also in renewing adult tissues and organs (this is the reason for adult stem cells). They can divide for much higher number of doublings and renew tissues and organs effectively. They

© Springer Nature Switzerland AG 2022
J.-Y. Yoon, *Tissue Engineering*, https://doi.org/10.1007/978-3-030-83696-2_5

play pivotal roles in reconstituting organ function, building tissue, and providing specialized cell types.

Zygote or *fertilized egg* can differentiate into all types of cells (including stem cells) in a body and eventually become a whole organism (e.g., human). It is called *totipotent stem cells* (toti = all, potent = ability). *Pluripotent stem cells* (pluri = plural) are similar to totipotent stem cells – they can differentiate into the cells found in all three germ layers, but not the extra-embryonic tissue such as placenta and umbilical cord. A good example is *embryonic stem cells* (*ES cells*) that can be found in the embryo. While pluripotent stem cells cannot become whole organisms, they are mostly sufficient for tissue engineering applications. *Multipotent stem cells* can differentiate into a limited number of cell types. A good example is hematopoietic stem cells, which can differentiate into all types of blood cells (red blood cells, white blood cells, and platelets) but not the other types of cells (Fig. 5.1). (Platelets are not cells, but they do originate from hematopoietic stem cells.

Both pluripotent and multipotent stem cells can be cultured in vitro. Much research has been conducted (and still being conducted) using this in vitro stem cell culture to understand the mechanism of their differentiation, prevention, and eventually treatment of congenital disabilities. Various drugs have been developed and tested on this in vitro stem cell culture to treat many diseases associated with stem cell differentiation or facilitate stem cell differentiation toward tissue or organ renewal.

Stem cells can be divided to increase their number, called *self-renewal*. Stem cells are differentiated under certain conditions, usually triggered by a particular gene, which we will discuss later in this chapter.

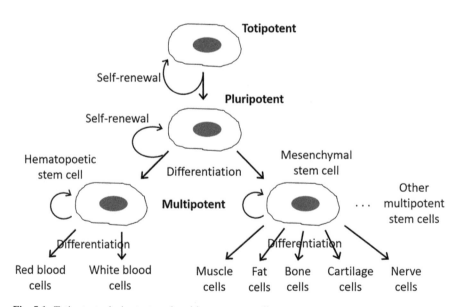

Fig. 5.1 Totipotent, pluripotent, and multipotent stem cells

5.2 Why Do We Need Stem Cells for Tissue Engineering?

Stem cells are also crucial for tissue engineering applications. Whether we build a tissue-engineered organ mimic or a tissue-engineered transplant, it is essential to secure healthy somatic cells. As many somatic cells' doubling time is quite long (the most notorious example is liver cells whose doubling time is a few hundred days), we also need a sufficient number of such somatic cells. In addition, there is a limitation in the number of possible doublings, called *Hayflick limit*. In many cases, unfortunately, an adequate number of healthy somatic cells is not always available. There usually are very few healthy somatic liver cells left in their liver for the patients who need tissue-engineered liver transplants. Most of such patients are in terminal conditions. The somatic cells from other healthy donors may be used for such a tissue-engineered transplant (this is still different from organ transplantation from a donor, where a whole organ is transplanted from a donor). However, since such somatic cells are not the same as the patient's own, compatibility issues and potential rejection should be considered and resolved.

The right solution would be:

(1) To obtain the adult stem cells from the patient's own body
(2) To proliferate them in vitro cell culture to produce a sufficient number of cells
(3) To seed them on a tissue-engineered scaffold and differentiate them into a specific cell type (with a particular function)
(4) To finalize the tissue-engineered organ mimic or tissue-engineered transplant

Keys to successful tissue-engineered organ mimic or transplant are (1) engineering design of scaffold that can provide the appropriate environment for stem cells' proliferation and differentiation, and (2) application of bioactive factors (physical and chemical) that can induce adequate differentiation to stem cells. Successful differentiation of stem cells leads to the self-organization of differentiated cells, optimum synthesis of extracellular matrix (ECM), and structural remodeling that can lead to successful tissue development and, subsequently, organ or organ-like structure.

5.3 Embryonic Versus Adult Stem Cells

Embryonic stem cells (*ES cells*) are typically obtained from a *blastocyst*, for example, 3–5 days post-fertilized egg (*zygote*) of mammals (including humans). Embryonic stem cells generally have been obtained from *in vitro fertilization* (*IVF*) for couples experiencing *infertility* (difficulty to reproduce naturally) (Fig. 5.2). A hormone is daily injected subcutaneously to the female to induce *ovarian hyperstimulation*. This daily injection produces an abnormally high number of eggs (several to 20 eggs compared to one egg per month). These eggs are extracted using an ultrasound-guided long needle. Sperm is also collected from a male and typically

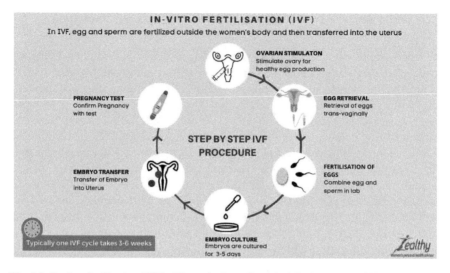

Fig. 5.2   In vitro fertilization (IVF). (Picture by https://zealthy.in/en in August 2020 and placed in the public domain. Accessed May 2021 from https://commons.wikimedia.org/wiki/File:In_Vitro_Fertilization_(IVF)_-_English.png)

centrifuged to isolate healthy ones. Egg and sperm are mixed and cultured in vitro (hence in vitro fertilization). Successful embryos are selected through microscopic observation and inserted into the uterus of a female. Additional hormones are daily injected to increase the chance of successful pregnancy. The number of embryos that can be introduced at a single attempt can be limited to two (e.g., in the UK) or not limited (e.g., in the USA), although 2–3 is typically recommended. Thus, there is a high chance for nonidentical twins or multiples, especially when a higher number of embryos are introduced.

For healthy females, as high as 20 eggs can be produced by ovarian hyperstimulation. Assuming half of them are successfully developed, and three are transferred to the female's uterus, seven healthy embryos will be left. These leftover embryos will be deep-frozen (*cryopreservation*). If the first embryo transfer attempt fails, a part of the frozen embryos will be thawed and transferred to the uterus again. Most likely, however, there will be leftover frozen embryos that will never be used. With the couple's and/or female's *informed consent*, these spare embryos can be used as a source for embryonic stem cells (ES cells).

As embryonic stem cells are obtained from embryos, they can differentiate into any cell types, thus pluripotent. Also, embryonic stem cells grow readily in different in vitro conditions. ES cells are susceptible to environmental conditions (temperature, pH, media, etc.). There are also legal and ethical issues with human ES cells, which will be discussed later.

On the other hand, adult stem cells can be obtained from adult mammals (including humans). There are much fewer legal and ethical issues in collecting and using adult stem cells from humans. However, they can differentiate into only a small number of cell types. For example, hematopoietic stem cells can only differentiate

Fig. 5.3 Differentiation of mesenchymal stem cells (MSCs)

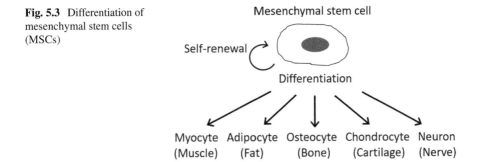

into red blood cells, white blood cells, and platelets. Most adult stem cells are multipotent but not pluripotent. They are more challenging to grow under in vitro conditions than embryonic stem cells. Besides, they are challenging to locate in a mammal (or a human body).

There are many types of adult stem cells available in human adults. In addition to hematopoietic stem cells described above, the following can be found – intestinal stem cells, neuronal stem cells, neural crest stem cells, olfactory adult stem cells, mammary stem cells, testicular stem cells, etc. They can differentiate into only a limited number of cell types as their name indicates. Together with hematopoietic stem cells, they are often classified as *tissue-specific stem cells*.

One type of adult stem cells deserves special mention – mesenchymal stem cells (MSCs). They can be found in bone marrow (similar to hematopoietic stem cells). However, they can differentiate into a large number of cell types (although still not pluripotent), for example, osteoblasts (bone cells), chondrocytes (cartilage cells), myocytes (muscle cells), adipocytes (fat cells), etc. (Fig. 5.3). Therefore, MSCs are quite different from the other tissue-specific stem cells. They have popularly been utilized for tissue engineering applications in the past, especially as alternatives to embryonic stem cells, when ethical and legal concerns become problematic.

While most ES cells are immortal, that is, they can proliferate (divide) indefinitely, many adult stem cells are not. For example, MSCs can proliferate (divide) only up to 50–70 generations and thus have a Hayflick limit (briefly discussed in Chap. 2 and will be discussed in detail in Chap. 10). However, the adult stem cells' Hayflick limit is still higher than that of the somatic cells.

5.4   Use of Embryonic Stem Cells for Tissue Engineering Applications

Embryonic stem (ES) cells can be differentiated in vitro using chemical and/or mechanical cues. Once differentiated into appropriate somatic cells, they can be utilized for various applications, including tissue engineering. If the final application is a tissue-engineered transplant or a patient-specific tissue-engineered organ

Somatic body cell with desired genes

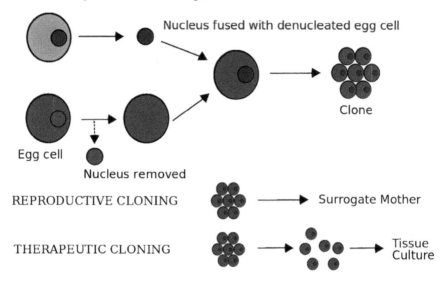

Fig. 5.4 Nuclear transfer methods for ES cells. (Picture by Belkorin and modified by Wikibob in November 2007 and placed in the public domain. Accessed May 2021 from https://commons. wikimedia.org/wiki/File:Cloning_diagram_english.svg)

mimic, the differentiated somatic cells must match patients precisely. However, as ES cells are sexually produced cells, there is no way to make these ES cells match those of a specific patient exactly. Therefore, several attempts have been made to "transfer" the patient's somatic cells' nucleus into an ES cell (Gurdon & Melton, 2008) (Fig. 5.4).

The first method is *somatic cell nuclear transfer (SCNT)* (Fig. 5.4). In this method, firstly, a nucleus of the patient's somatic cell is isolated. Secondly, an egg is harvested from a female donor, typically by ovarian hyperstimulation (refer to the IVF description in Sect. 5.3), and its nucleus is physically removed. Thirdly, the nucleus of a somatic cell is injected into the "empty" egg. Finally, mechanical stimuli are provided to induce embryonic development. However, correct embryonic development's success rate has been proved extremely low, leading to the need for a very high number of donated eggs and subsequently creating various ethical and legal concerns.

The second method is *cell fusion* (Fig. 5.5). Cell fusion can naturally occur in mammals (and humans), where cells of the same type are fused – called *homotypic cell fusion*. Merged nuclei form *synkaryon*. However, for stem cell applications, cells of different types are typically fused – called *heterotypic cell fusion*. Merged cell nuclei form *heterokaryon*. Gene activity will eventually be switched to that of a dominant cell (somatic cell).

Both SCNT and cell fusion are considered the method of reprogramming nuclei, that is, nuclear reprogramming. They are the methods of switching the gene

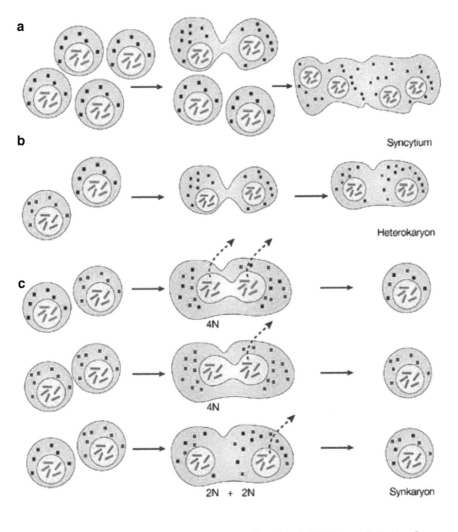

Fig 5.5 Cell fusion. (Ogle et al., 2005. Reprinted with permission, (C) 2005 Springer Nature)

expression of one kind, for example, embryonic stem cell, to another, unrelated type, for example, somatic cell. However, adding nuclei of desired cells (somatic cells) is not sufficient to induce successful gene expression and subsequently differentiation. Toward this, proteins (typically chromosomal proteins) from somatic cells are often added to assist the expression of somatic cells' genes.

5.5   Induced Pluripotent Stem Cells (iPSCs)

Many genes have been identified that can induce differentiation to stem cells. Takahashi and Yamanaka have introduced four such genes – Oct3/4, Sox2, c-Myc, and KLF4 – to adult mouse fibroblasts using a virus as a delivery vehicle (vector), for example, viral transfection. They reported in 2006 that some of these cells showed the appearance of characteristic embryonic stem cells. After further selection, these cells were shown to enter all cell lineages when transplanted to host embryos (Takahashi & Yamanaka, 2006) (Fig. 5.6). The identical phenomenon was also discovered with human cells, using Oct4, Sox3, Nanog, and lin28. This phenomenon was called *induced pluripotency,* and the resulting cells were called *induced pluripotent stem cells* (*iPSCs*).

Since this discovery, these procedures have been confirmed by many others and extended with various cell types. The resulting stem cells appeared to be the same as embryonic stem cells. With iPSCs, there is no need to worry about the legal and ethical issues of using embryonic stem cells and the availability issues of adult stem cells. Any somatic cells from the patient can be harvested, converted to iPSCs, differentiated, and eventually used for cell replacement therapies and tissue engineering applications.

Question 5.1 How are the induced pluripotent stem cells (iPSCs) made? What is the primary benefit of using iPSCs over embryonic stem cells (ESCs) toward tissue-engineered transplant for human subjects?

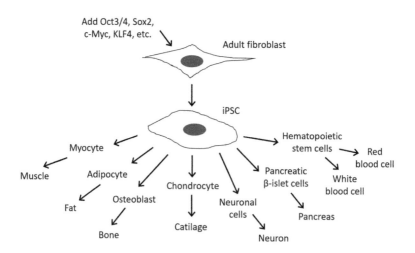

Fig. 5.6 Induced pluripotent stem cells (iPSCs)

5.6   Isolation of Stem Cells

Regardless of stem cells' source, we should isolate a best-behaving stem cell line. The best-behaving stem cell usually expresses a specific cell surface marker protein, which can be used for direct isolation of a particular stem cell line. Many cell surface markers have been identified and subsequently named, for example, c-Met, p75NTR, etc. However, there exist a large number of cell surface markers whose functions have not been clearly identified and subsequently not named. They are referred to as the *cluster of differentiation* (*CD*) proteins and numbered numerically, for example, CD1, CD2, CD3, etc. Close to 400 CDs for humans have been identified so far.

CD proteins can be utilized to isolate a specific stem cell line via negative or positive selection. Antibodies to a specific CD protein are immobilized to a solid support, typically beads or polymers packed within a chromatographic column. In negative selection, an antibody (or antibodies) is used to target the CD protein(s) that is present on nontarget cells (i.e., cells that you do not want). Cells expressing this nontarget CD protein(s) bind to the column, while the cells not expressing this nontarget CD protein will be collected downstream. In positive selection, the cells expressing the target CD protein are captured while other cells are washed away.

Let us take an example of blood cell isolation. Table 5.1 shows a partial list of CD proteins for blood cells. In the negative selection shown in Fig. 5.7, anti-CD45+ is immobilized on the solid support and captures all cells that express CD45+, that is, all types of blood cells. All other cells will be collected downstream. In the positive selection also shown in Fig. 5.7, anti-CD4+ is immobilized on the solid support and captures the cell expressing CD4+, that is, only T helper cells. All other cells will be washed away.

Question 5.2   To capture monocytes and macrophages, what antibodies should be used? Should you use negative or positive selection?

Question 5.3   To capture all types of T cells, what antibodies should be used? Should you use negative or positive selection?

Table 5.1   A partial list of CD proteins for blood cells

Blood cell types	CD proteins
Monocyte and macrophage	CD45+ CD14+
T cell	CD45+ CD3+
T helper cell	CD45+ CD3+ CD4+
Cytotoxic T cell	CD45+ CD3+ CD8+
B cell	CD45+ CD19+ or CD45+ CD20+
Natural killer cell	CD45+ CD16+ CD56+ CD3−
Thrombocyte (platelet)	CD45+ CD61+

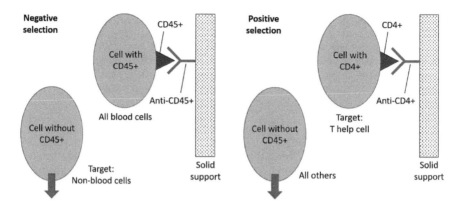

Fig. 5.7 Negative versus positive selection using CD proteins

5.7 Culturing Stem Cells

Scaffold Once the stem cell line is isolated, we need to provide a suitable environment for these cells to proliferate and differentiate in a manner we desire. For tissue engineering applications, a *scaffold* is required, as already explained in Chap. 1. A scaffold can be made in 2D or 3D. This scaffold can be modified to include a specific chemical or biological ligand (functional groups, peptides, enzymes, proteins, etc.) that can accommodate successful adhesion of stem cells and potentially signal its differentiation. It can also hold biological molecules (e.g., differentiation factors) within its structure and release them in response to the microenvironment so that specific genes are expressed to induced differentiation.

Co-culture Environment Many stem cells, especially embryonic stem cells, require a *co-culture environment* as they do not provide necessary growth factors, soluble and insoluble factors, and some morphogens. The added layer of cells is called a *feeder layer*. For embryonic stem cells, embryonic fibroblasts are typically used as a feeder layer. Fig. 5.8 (top) shows the multilayered co-culture of stem cells, where stem cells are added to form the first layer and chondrocytes are added on top of it to create the second layer. Both layers are sequentially cross-linked to confine the cells within the cross-linked gels, while smaller molecules are free to move across two layers. Figure 5.8 (bottom) shows the more recent co-culture of stem cells, utilizing a *transwell insert*. The insert is a filter cup with a semipermeable membrane at the bottom. This filter cup is inserted into a larger tissue culture flask or a smaller microwell plate, where the stem cells are cultured on the filter cup, and the feeder cells (e.g., fibroblasts) are cultured in the bottom tissue culture plate. Cells cannot move across the membrane, while small molecules are free to move across the semipermeable membrane.

Multilayer

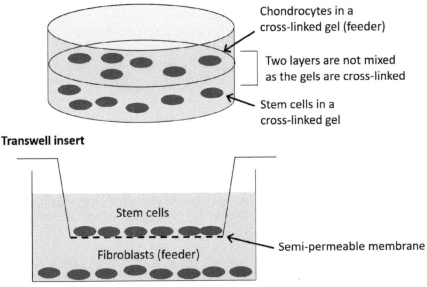

Transswell insert

Fig. 5.8 Top: multilayered co-culture of stem cells. Each layer contains gels and cross-linked (using UV light) to maintain the integrity of each layer and to prevent mix-up between two layers. Bottom: a transwell insert on a microwell plate is used to create a stem cell co-culture

Bioreactor Requirements Stem cell culture (co-culture) requires a low shear environment to prevent any shear-induced damage to stem cells and prevent mix-up of two layers (for co-culture). However, you should provide good mass transfer of oxygen and nutrients between cells and two layers for its proliferation (self-renewal) and differentiation. A fluidic connection between stem cell and feeder layer and between the cell layers and media may be necessary, which will be further discussed in Chap. 10.

Question 5.4 A tissue-engineered skin transplant is being developed in a bioreactor using an appropriate polymeric scaffold. Two different cell types are being co-cultured in a bioreactor:

	Doubling time (t_d)	Number of cells in the skin
Keratinocytes (makes up epidermis)	66 h	60%
Fibroblasts (makes up dermis)	21 h	40%

The bioreactor will be operated for 20 days, by when the overall cell density of 5×10^7 cells/mL and the total volume of 100 mL (not counting the volume of a polymeric scaffold) will be reached. Calculate the number of keratinocytes and fibroblasts that need to be initially seeded onto the polymeric scaffold. Both cells follow the first-order growth kinetics.

Hint 1. Calculate the final cell number (N) using the final cell density and the cell fractions in the skin.

Hint 2. Use the equations shown in Chap. 3 but replace X and X_0 with N and N_0 as $X = N/V$ and V is constant. Using the given t_d and $t = 20$ days, calculate N_0 for both cell types.

Successful differentiation of embryonic stem cells can typically be assessed by checking *embryoid body* (*EB*). Embryoid bodies are essentially sphere-shaped clusters of stem cells (Fig. 5.9). Embryoid bodies can also be found in other types of stem cells, for example, those from SCNT, cell fusion, as well as iPSCs. The presence of EBs signals the differentiation of stem cells and the morphogenesis for embryonic development. Differentiation is typically characterized by the expression of the gene *Oct4* (of course, this is not the only gene expressed for differentiation). Oct4 has already been mentioned in the iPSC section.

5.8   Morphogenetic Factors

During embryonic development or *embryogenesis*, stem cells are differentiated into many different cell types and develop into specific tissues and organs. Such a process is called *morphogenesis*. Morphogenesis is controlled by many factors, including physical cues (mechanical stimuli, shape and morphology of surrounding

Fig. 5.9 Embryoid bodies from suspension cell culture. (Picture taken by Stemcellscientist in April 2012 and placed in public domain. Accessed February 2021 from https://commons. wikimedia.org/wiki/ File:MESC_EBs.jpg)

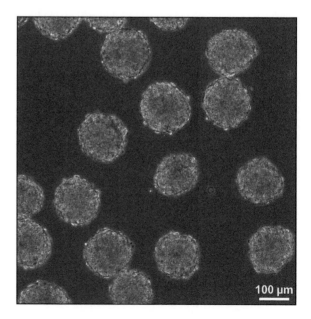

100 µm

matrix, etc.) and biological factors (growth factors, differentiation factors, etc.). These factors (physical cues and biological factors) are collectively called *morphogenetic factors*. Therefore, it is a broader term than differentiation factors, although focused more on embryonic development. Precise control of physical cues and controlled release of biological factors from a scaffold can provide instructive signals to the stem cells for their successful differentiation and embryogenesis-like tissue/organ development.

Here is a list of well-known morphogenetic factors that can be used toward stem cell-based tissue engineering applications:

Bone Morphogenetic Protein 2 (BMP-2)   BMP-2 is one example of bone morphogenetic proteins. As its name suggests, it plays a vital role in the development of bones and cartilages. Adding BMP-2 to the in vitro culture of embryonic stem cells induces the formation of embryoid bodies.

Tissue Growth Factor β (TGF-β)   TGF-β has multiple roles, mostly focused on cell growth and proliferation as its name suggests, but also plays essential roles in differentiation and apoptosis (programmed cell death). In fact, BMP-2 also belongs to the TGF-β family of proteins.

Extracellular Matrix (ECM) Microenvironment   Extracellular matrix (ECM) can also provide physical cues to the stem cells' proliferation and differentiation. The main backbone of ECM is collagen fibers. And these collagen fibers are interspersed with much thinner polymer chains – *glycosaminoglycans (GAGs)*. GAGs are long linear polymers of sugars (glucose), or polysaccharides, modified with amino acids. Most commonly found GAGs include heparin sulfate (HS), chondroitin sulfate (CS), keratan sulfate (KS), and hyaluronic acid (HA). For example, there have been attempts of duplicating collagen fibers with *polyethylene glycol (PEG)*-based hydrogels, where the network of PEG polymers is cross-linked to form a network that contains a large volume of water. Mammalian cells adhere to this hydrogel and proliferate quite well. When adding one of the above-listed GAGs to this PEG hydrogel, stem cells start to differentiate, clearly indicating its ability to induce differentiation. A scaffold can be fabricated or patterned like the ECM microenvironment toward inducing such differentiation and morphogenesis. This topic will be further covered in Chaps. 8 and 9.

5.9   Hazards of Stem Cell Differentiation

Successful differentiation of stem cells is essential and must be checked before their use in tissue engineering applications, especially for transplantation. Injection of stem cells directly into a patient will cause serious side effects, as we may not control their differentiation in a well-orchestrated manner. Such an act typically leads to cancer formation made of different types of cells or tissue, called a *teratoma*.

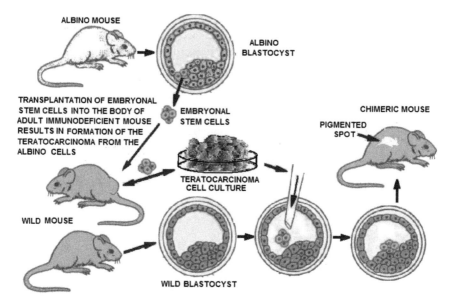

Fig. 5.10 Direct injection of embryonic stem cells into a mouse typically leads to teratoma formation. (Picture was taken by Dmitry Dzhagarov in February 2014 and place in the public domain. Accessed February 2021 from https://commons.wikimedia.org/wiki/File:Healthy_chimeric_mouse_from_teratocarcinoma_cells.JPG)

Occasionally, a separate term, *teratocarcinoma,* refers explicitly to the tumor tissue from germ cells during embryonic development, whereas teratoma can be found in adult tissues (Fig. 5.10).

We should precisely manipulate the culture conditions if we want to avoid teratoma formation. Besides, isolation of best-behaving stem cells is necessary. Despite all these efforts, the differentiated stem cells may still behave somewhat differently from healthy somatic cells.

5.10   Laboratory Task 1: Culturing Stem Cells

In this task, we will culture human iPSCs (hiPSCs). One of the popularly used media for culturing hiPSCs is the *mTeSR1 medium* (Fig. 5.11). This protocol is a modified version that has been published by Pistollato et al. (2017).

1. Prepare a tissue culture flask (TCP) or a petri dish and coat it with the basement membrane matrix that is qualified to cultivate human embryonic stem cells. *The basement membrane* is a type of extracellular matrix where the anchorage-dependent mammalian cells adhere to and proliferate quite well. Matrigel is the most well-known example (Fig. 5.10).

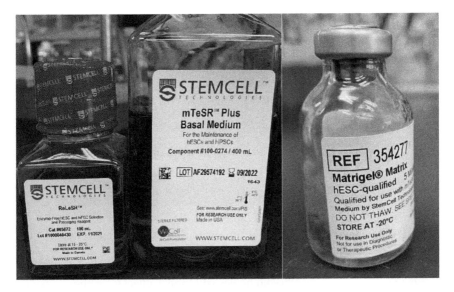

Fig. 5.11 ReLeSR™, hESC, and hiPSC selection and passaging reagent (left); mTeSR™ medium for hESC and hiPSC (middle); Matrigel® for coating TCP (right)

2. Dilute 200 µL of basement membrane matrix in 20 mL of DEME/F12 medium (1:100 dilution).
3. Coat the TCP this diluted matrix (5 mL per dish). Incubate for 1 h at 37 °C. Note: Matrigel-coated TCP is commercially available, and you can skip steps 1–3.
4. Thaw hiPSC colonies and plate them in the above matrix-coated TCP at around 100 fragments per dish.
5. Incubate at 37 °C and 5% CO_2 using a CO_2 incubator.
6. Change medium daily.
7. Passaging: Detach the colony fragments from the surface using a pipette. As they are undifferentiated and rounded, they can be easily detached without using an enzyme. Transfer the detached colonies into the new matrix-coated TCP using steps 4 and 5.
8. Image the cells using the protocol described in Chap. 4 (no fluorescent staining necessary). Undifferentiated hiPSCs should be round in shape with large nucleoli and without abundant cytoplasm (Fig. 5.12). They are also flat and tight-packed. Perform passaging as necessary based on the imaging results.

5.11   Laboratory Task 2: Embryoid Body Formation

9. Detach the hiPSC fragments using the same method described in step 7. Centrifuge the cells to collect just the cells (optional). Resuspend them in 5 mL of hiPSC medium.

Fig. 5.12 iPSCs are round, tightly packed cells with relatively large nuclei. They form circular colonies

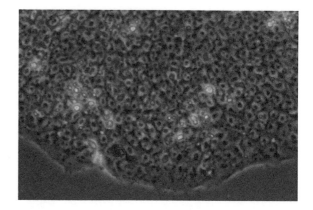

10. Plate the fragments in an ultra-low attachment TCP (or petri dish), that is, without basement membrane matrix.
11. Incubate at 37 °C and 5% CO_2 using a CO_2 incubator.
12. On the next day, some embryoid bodies will form. Collect the embryoid bodies using a pipette.
13. Centrifuge them at $112 \times g$ for 1 min. Remove supernatants. Resuspend them in 5 mL of hiPSC medium.
14. Image the embryoid bodies using the protocol described in Chap. 4 (no fluorescent staining necessary).
15. Repeat steps 10–13 until a good number of embryoid bodies are obtained.

The obtained embryoid bodies can then be replated back to the basement membrane matrix (Matrigel). Instead of using a standard culture medium for stem cells, the *neuroepithelial induction medium* can be used to differentiate hiPSCs into neuroepithelial aggregates. This process will take up to one month and is quite a delicate process requiring precision and accuracy (Fig. 5.13).

Review Questions
1. Compare totipotent, pluripotent, and multipotent stem cells. List examples of all three categories.
2. What are the ethical issues in using human embryonic stem cells (hESCs)?
3. What is a mesenchymal stem cell? What are the advantages of mesenchymal stem cells over embryonic stem cells?
4. Describe somatic cell nuclear transfer (SCNT). Compare it with cell fusion.
5. What is chromosomal protein exchange, and when is it needed?
6. What is inducible pluripotency?
7. What is CD protein?
8. Compare negative versus positive selection in isolating cell line.
9. Why does stem cell culture require co-culture? In other words, why do you need a feeder layer?
10. How can the stem cells be differentiated in vitro? In other words, what are morphogenetic factors, and what do they do?

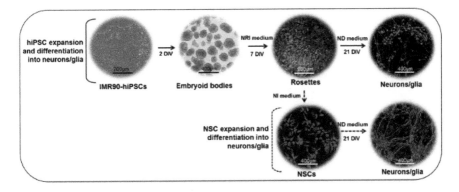

Fig. 5.13 Differentiation of human iPSCs into neurons. (Pistollato et al., 2017 (C) Open access article distributed under the terms of the Creative Commons Attribution 4.0 International License)

11. What is a bone morphogenetic protein (BMP)? What is a transforming growth factor (TGF)?
12. What is an embryoid body, and why is it essential for stem cells?
13. What is teratoma? How can you prevent teratoma formation?

References

Gurdon, J. B., & Melton, D. A. (2008). Nuclear reprogramming in cells. *Science, 322*(5909), 1811–1815. https://doi.org/10.1126/science.1160810

Ogle, B. M., Cascalho, M., & Platt, J. L. (2005). Biological implications of cell fusion. *Nature Reviews Molecular Cell Biology, 6,* 567–575. https://doi.org/10.1038/nrm1678

Pistollato, F., Canovas-Jorda, D., Zagoura, D., & Price, A. (2017). Protocol for the differentiation of human induced pluripotent stem cells into mixed cultured of neurons and glia for neurotoxicity testing. *Journal of Visualized Experiments, 124,* e55702. https://doi.org/10.3791/55702

Takahashi, K., & Yamanaka, S. (2006). Induction of pluripotent stem cells from mouse embryonic and adult fibroblast cultures by defined factors. *Cell, 126*(4), 663–676. https://doi.org/10.1016/j.cell.2006.07.024

Chapter 6
Biomaterial Surfaces

The previous chapters' foci were cells, that is, cell culture, cell metabolism, cell imaging, and stem cells. However, in tissue engineering applications, the surfaces are equally important that these cells adhere to and grow on. These surfaces are made from biomaterials, which is the primary learning objective of this chapter. Such biomaterial surfaces also make up the scaffold, typically in 3D. However, in this chapter, we will limit the discussion only to the surface properties and surface modifications of biomaterials, and we will discuss the scaffold fabrication in the later chapters.

Inquiry 1. Why and when do you need the engineering of biomaterial surfaces for tissue engineering applications (including tissue-engineered organ mimic and tissue-engineered transplant)?

Inquiry 2. What are the biomaterial surfaces' requirements for tissue engineering applications (including tissue-engineered organ mimic and tissue-engineered transplant)?

6.1 Development of Biomaterial Surfaces for Tissue Engineering

Whether your final goal is a tissue-engineered organ mimic or a tissue-engineered transplant, you cannot expect your cells (somatic or stem cells) to self-assemble by themselves and magically create a fully functioning organ with appropriate structure. As most tissue-engineered devices utilize anchorage-dependent cells, you will need a surface that can mimic the natural extracellular microenvironment, including the *extracellular matrix (ECM)* and ECM-bound growth factors.

As shown in Fig. 6.1, ECM is essentially a gel. Protein fibers provide a polymeric network, glycosaminoglycans (GAGs) interweave in between, and water fills

© Springer Nature Switzerland AG 2022
J.-Y. Yoon, *Tissue Engineering*, https://doi.org/10.1007/978-3-030-83696-2_6

Fig. 6.1 Mammalian cell (bottom) is interacting with extracellular matrix (ECM; top). Vitronectins and laminins (not shown) can also serve as glue proteins

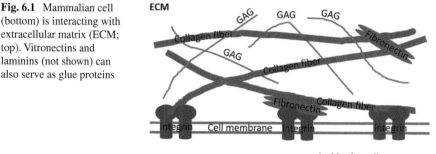

the remaining structure. GAGs are polysaccharides modified with amino acids; refer to Sect. 5.8. A significant portion of ECM is water. A tiny amount of ECM is solid. The primary protein fibers are collagens (green-colored in Fig. 6.1), acting as the primary backbones of ECM. Occasionally, collagen fibers are replaced with *elastins*, where the tissue requires elasticity. GAGs are often coupled with proteins, known as *proteoglycans*. Collagen fibers (occasionally elastins) provide resistance to tensile forces, while the hydrated networks provide resistance to compressive forces. Fibronectins, vitronectins, and laminins are "glue" proteins that connect mammalian cells to ECM structures. Integrins in the cell membrane (phospholipid bilayer) can recognize collagen fibers, fibronectins, vitronectins, and laminins, which form focal adhesion (discussed in the next chapter).

ECM should provide mechanical support, guidance cues for cellular migration, and 3D network structure allowing cell–cell communications (Fig. 6.1). Cell–cell communications are achieved by growth factors, *cytokines* (a broad category of small proteins responsible for cell–cell communications), etc. It is desirable to recapitulate tissue's and organ's developmental processes, especially when using stem cells. These include differentiation and morphogenesis.

Toward this goal, we can utilize an artificial biomaterial surface. While simplified biomaterial surfaces are easy to fabricate and use, oversimplification typically leads to the failure in appropriate tissue (and eventually organ) development. On the other hand, the exact recapitulation of natural ECM is quite challenging to accomplish. This chapter will limit our discussion to the choices of biomaterial surfaces and their surface modification methods.

6.2   Size and Shape Requirements

2D Versus 3D Biomaterial surfaces can be offered in 2D, that is, flat surfaces, or in 3D, that is, network of cross-linked polymers (e.g., gels) or porous materials. 2D surfaces are easy to fabricate and subsequently used quite commonly. Media exchange (feeding) and passaging are also easy, which also make the scale-up easy. However, cells tend to "flatten" on 2D surfaces, leading to inferior cytoskeleton development (Fig. 6.2). 3D surfaces better recapitulate the natural ECM's and cells' morphology, and phenotypes are closer to those in vivo. Certain functions of cells cannot be demonstrated on 2D surfaces – in that case, 3D surfaces must be used (Fig. 6.2). However, feeding may be difficult. Passaging is even more challenging – how can you harvest your cells (and remove dead or poorly performing cells) from gels or porous matrices?

Length Scale To provide better surface anchorage, micrometer structures are often "patterned" on biomaterial surfaces. For example, micrometer fibers can be patterned on a biomaterial surface to mimic the collagen fibers in ECM. Large-sized fibers, for example, 10 μm, are easy to fabricate and be patterned on a biomaterial surface. However, typical mammalian cells' sizes are on a similar scale, leading to the nonoptimum anchorage of cells (Fig. 6.3). They can be flattened out on the surface and/or with minimum focal adhesion points (the latter will be further discussed in the next chapter). Small-sized fibers, for example, 0.1 μm, are always better at the cost of difficulty in fabrication and patterning (Fig. 6.3).

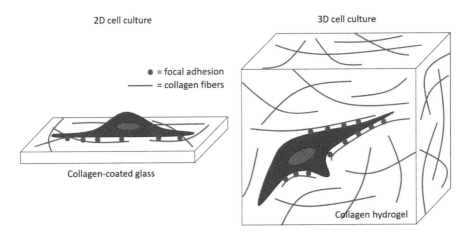

Fig. 6.2 Cells on a 2D surface (collagen-coated glass) and a 3D gel (collagen gel). (Baker BM, Chen CS. 2012. Deconstructing the third dimension – how 3D culture microenvironments alter cellular cues)

Fig. 6.3 Effect of length scale on cellular adhesion

Fibroblast on 10 μm fibers   Fibroblast on 0.1 μm fibers

Fig. 6.4 PE, PTFE, and PDMS

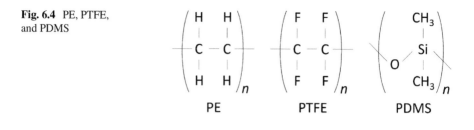

PE   PTFE   PDMS

6.3   Synthetic Materials

Like other biomaterials, biomaterials for tissue engineering applications can be categorized into two: synthetic versus natural. Synthetic materials include polymers, such as polyethylene (PE), polytetrafluoroethylene (PTFE), and polydimethylsiloxane (PDMS) (Fig. 6.4).

Polyethylene (PE) is quite a simple hydrocarbon polymer, essentially an extremely long alkane. While the starting monomer is ethylene (belongs to alkene; hence the name polyethylene), one of its double-bond is opened to form a long polymer consisting only of single bonds (thus alkane). Through creating branches and connecting them (cross-linking), desirable material properties can be achieved, ranging from soft and elastic materials (appropriate for soft tissues) to hard and rigid materials (suitable for hard tissues such as bones). However, as PE is hydrocarbon without any polar or charged components, it is strongly hydrophobic and not very compatible with ECM. Water is still the primary component in both cells and ECM.

Polytetrafluoroethylene (PTFE) and its derivatives (including *Teflon*®) are essentially hydrocarbon polymers where hydrogen atoms are replaced with fluorides. Teflon has primarily been used as nonstick coatings for various applications (including cookware such as pans) due to its extreme hydrophobicity and porosity. It can also be used for cardiovascular applications due to its "nonstick" nature, despite its obvious hydrophobicity and incompatibility with cells and ECM.

Polydimethylsiloxane (PDMS) has recently become popular as a prime material for fabricating *lab-on-a-chip* (*LOC*) devices through various soft lithography techniques. *Soft lithography* allows the fabrication of various structures (fibers, channels, posts, wells, etc.) in micrometer scales similar to the conventional photolithography techniques. Also, it does not require complicated fabrication procedures and minimizes the use of a clean lab facility. Such microfabrication capability has also been adopted for modifying the surfaces for various biomaterial and tissue engineering applications (refer to Fig. 6.2). However, PDMS is also hydrophobic. The polar oxygen molecule in the backbone is "buried" by the methyl groups and not exposed to the outside, rendering it less biocompatible like PE.

Membranes and filters are made from various polymers, and commonly used as biomaterial and tissue engineering applications (*membranes* are typically the filters with smaller pore sizes). Many ECMs have membrane structures, the most famous example being basement membrane, which will be discussed in the next section.

Oligopeptides can also be used for biomaterial and tissue engineering applications. Proteins are essentially the polymers of amino acids (there are 20 different types), called *polypeptides*. *Oligopeptides* are short sequences of peptides and can easily be synthesized chemically. These days, many commercial vendors can synthesize and sell the oligopeptides at your design. These oligopeptides can then be self-assembled to create a bigger structure that can mimic the structural components of ECM, or they can be used as a coating material to synthetic polymers to enhance cellular adhesion.

There are many advantages of using synthetic materials as tissue-engineered scaffolds. These include:

- Ability to control the initial shape and geometry of the scaffold.
- Mechanical support with precise properties.
- Ability to control cell adhesion precisely.
- Ability to deliver growth factors and other bioactive compounds from the scaffold.

Disadvantages include:

- Challenges in design and matching the exact components of ECM.
- Likely interference from nearby ECM in the long term – this leads to the reorganization of the scaffold and nearby ECM, affecting its mechanical properties.

To overcome these advantages, we may want to use natural materials, discussed in the following section.

6.4 Natural Materials

Natural materials derived from living organisms can be used as biomaterial surfaces. For biomaterial and tissue engineering applications, both polysaccharides (polymers of various sugars) and proteins (polypeptides) have been used.

Polysaccharides for tissue engineering applications include alginate, dextran, chitosan, cellulose, starch, and hyaluronan (= hyaluronic acid). Figure 6.5 shows their chemical structures, indicating their structural similarities. *Alginate* (= alginic acid) is a polysaccharide derived from sea algae (hence the name alginate). It is a block copolymer of D-mannuronic acid and L-guluronic acid. *Dextran* is derived from bacteria. There is an additional covalent bonding point (dashed line) in its structure, making the cross-linking process easier. *Cellulose* is the major component of plant cell walls and consequently of paper. It is easy to modify chemically. Specifically, papers are easy to fabricate (through bending, cutting, coating, etc.). However, cellulose is less cell-friendly, and its application has been limited to orthopedic applications. *Starch* is also derived from plants. Unlike cellulose, starch can be made into a biodegradable substrate (used for biodegradable plastic bags). However, starch is somewhat less popular than synthetic biodegradable polymers – PLA and its derivatives, discussed in Sect. 6.9.

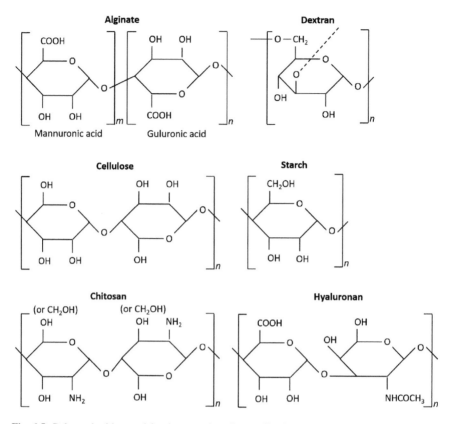

Fig. 6.5 Polysaccharides used for tissue engineering applications

Chitosan is a polysaccharide derived from marine crustaceans (crabs, lobsters, shrimps, etc.) and fungi's cell walls. Chitosan has two amine groups in its repeating units, not found in the other polysaccharides listed above. Glycosaminoglycans (GAGs) in ECM are also polysaccharides modified with amine groups. Therefore, there is a substantial similarity between chitosan and GAG, and it has popularly been used for biomaterial and tissue engineering applications.

Hyaluronan or *hyaluronic acid* (*HA*) is one type of GAGs and has popularly been used for biomaterial and tissue engineering applications. GAGs are classified into four groups: (1) heparin (= heparan sulfate or HS), (2) chondroitin sulfate (CS), (3) keratan sulfate (KS), and (4) hyaluronan (= hyaluronic acid or HA). As their names indicate, hyaluronan (HA) is the only one not sulfated.

All these polysaccharides can be cross-linked to form a network. Once filled with water, they form a hydrogel, which is biocompatible and cell friendly. We will further discuss hydrogels in Sect. 6.8.

Proteins are more popular natural materials for tissue engineering applications. Specifically, the purified ECM components are commonly used, including collagen (the primary backbone of ECM), elastin (the primary backbone of ECM from "elastic" tissues), fibronectin (tissue glue), etc. Among these, *collagen* is the most popularly used for tissue engineering applications. Collagens are produced by *fibroblasts* (frequently found in most connective tissues) and *osteoblasts* (bone cells). In other words, collagens are found in both soft and hard tissues. Collagens initially have a triple helix structure with random structures at both ends. This precursor form is called *procollagen*. Through enzymatic cleavage, such random structures are removed to form proper collagen, called *tropocollagen*, shown in Fig. 6.6 (1.5 nm diameter and ~ 300 nm long). These collagens are bundled together to form *collagen fibrils* (10–300 nm diameter), as shown in Fig. 6.6. Note that the collages are quarter-staggered to provide necessary mechanical properties. Collagen fibrils are further cross-linked together to form *collagen fibers* (1–20 μm), as shown in Fig. 6.6. There are many different types of collagens (> 20 types), where collagen type I is identified the first and most commonly used. *Collagen type I from rat tail* (= *rat tail collagen I*) is widely used collagen for biomaterial and tissue engineering applications. Other collagen types can be produced by recombinant technology by inserting genes that can make a specific type of collagen into bacteria, yeast, etc.

Another natural material is very popularly used for biomaterial and tissue engineering applications – perhaps more popular than collagen these days. It is *Matrigel®* from BD Biosciences. It is solubilized *basement membrane* extracted from mouse tumors, typically enriched with laminin. Laminin is another type of tissue glue (the other being fibronectin), while it is more commonly found in the basement membrane. A basement membrane is a type of ECM whose shape is characterized by a thin, membrane-like sheet structure. Cells are anchored to one side of

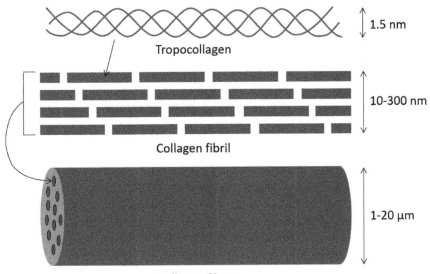

Fig. 6.6 Tropocollagen (top), collagen fibril (middle), and collagen fiber (bottom)

the basement membrane, which provides structural support to the cells and allows cell signaling molecules to be transported.

6.5   Nonspecific Cell–Surface Interactions

Cells can interact with biomaterial surfaces in many ways. We can categorize it into two – nonspecific and specific interactions. In nonspecific interactions, cells can interact with the biomaterial surface by the following:

– *Electrostatic interaction*: Sufficiently strong charges on the cell surface and the biomaterial surface can generate electrostatic interactions between cell and biomaterial. It includes both *electrostatic attraction* and *electrostatic repulsion*. Recall that the cells' membranes are made from phospholipid bilayers, and their head groups are negatively charged by phosphate groups. The negatively charged biomaterial surfaces will repel the cells via electrostatic repulsion, and the positively charged biomaterial surface will attract the cells via electrostatic attraction.

– *Hydrogen bonding:* Hydrogen bonding is sometimes considered a special type of electrostatic interaction, where negative and positive polarities (known as *dipoles*) from two different molecules are associated together. In a narrower sense, electrostatic attraction occurs between two differently charged molecules, while hydrogen bond forms between two different uncharged polarities (diploes). All cells have membrane proteins on their surfaces. Proteins have positive/negative charges (from positively or negatively charged amino acid residues) or positive/negative dipoles. Their *isoelectric point* (*pI*) determines the net charges of mammalian proteins, where pI is the pH where the net charge of a protein becomes zero. As the pI's of most proteins are slightly lower than physiological pH (7.2–7.4), the net charges of most proteins are slightly negative. Many mammalian cells also have *glycoproteins* on their surfaces, usually uncharged but polar with negative dipoles due to the polysaccharide presence. (Note: glycoproteins are proteins modified with sugars. Proteoglycan is a subset of glycoproteins, where the sugar portion is polysaccharides, and often used interchangeably with GAGs.) Therefore, the biomaterial surfaces with positive charges or positive dipoles would allow nonspecific cellular adhesion. However, as such charges and dipoles are not evenly distributed on their surfaces, proteins can easily change their orientation to accommodate their adhesion on the surfaces; hence, the net polarity does not matter. Nonspecific cellular adhesion is preferred on the surfaces with any charges or any dipoles, that is, *hydrophilic* (hydro = water, philic = loving) surfaces.

– *Hydrophobic interactions:* Hydrophobic interactions refer to the association of two molecules or functional groups that are hydrophobic (hydro = water, phonic = hating). They occur only in aqueous media because water molecules, being very polar solvent, try to exclude such hydrophobic molecules/groups from water solvent. Cells' membrane surface is strongly negatively charged and thus not hydrophobic. Glycoproteins are also not hydrophobic due to the sugar presence. However, some membrane proteins may be partially hydrophobic, depending on how many of their amino acid residues are hydrophobic. However, this is not a favored way for inducing cell–biomaterial interaction. The overall strength of cell adhesion on hydrophobic biomaterial is relatively weak due to the lack of hydrophobicity in the cell membrane and glycoproteins. If a specific membrane protein is associated with a hydrophobic surface, they usually undergo a *conformational change*. Proteins are "folded" by hydrophobic interaction to make their particular shape and function appropriately through hiding hydrophobic amino acid residues to their core and exposing hydrophilic residues to their surface. Once membrane proteins are associated with a hydrophobic biomaterial surface, there is no need to "hide" such hydrophobic amino acid residue, thus changing their shape (conformational change) and subsequently losing their functions.

– *Van der Waals interactions:* This is a weaker version of electrostatic interactions. Like electrostatic interactions, there are van der Waals attraction and van der Waals repulsion. In aqueous media (like ECM), however, van der Waals interactions are substantially weaker than the other three mentioned above.

6.6 Specific Cell–Surface Interactions

Cells can also adhere to the ECM and biomaterial surfaces in a specific manner. Such specific cell–surface interactions are always mediated by cell surface receptors (= *cell receptors*). The most famous example is *integrin*, which recognizes many different proteins (collagen, elastin, fibronectin, vitronectin, and laminin) in the ECM. Specifically, integrin-mediated cellular adhesion is called focal adhesion, further discussed in the next chapter.

6.7 Bone Biomaterials, Apatite, and Bioglass

Bones are quite different from other tissues and organs – the significant difference is the presence of many minerals, specifically apatite. An *apatite* is a group of phosphate minerals, specifically calcium phosphate $Ca_5(PO_4)_3^+$ for mammals. They can bind to hydroxyl group to form hydroxyapatite $Ca_5(PO_4)_3OH$, fluoride ion to form fluorapatite $Ca_5(PO_4)_3F$, or chlorine ion to form chlorapatite $Ca_5(PO_4)_3Cl$. While calcium phosphates are the main form of minerals in mammals, calcium carbonates can also be found in, for example, clams and corrals. These minerals are combined with collagens to yield very hard and stiff mechanical properties to bones and teeth.

Bones have traditionally been replaced wholly by metal implants. These metals include cobalt-based alloys (e.g., CoMoCr), titanium, titanium-based alloys, etc. Ceramics (including alumina, titania, zirconia, etc.) and engineering plastics (polyurethane = PU, polyethylene = PE, etc.) are often used to replace bones. However, the material properties of engineering plastics are often inferior to metals, leading to limited applications. The major drawback of these bone implants is their poor osseointegration (osseo or osteo = bone) – they do not integrate very well into the nearby tissues (other bones, cartilages, etc.).

We can create a micron-scale (< 10 μm) roughness on the metallic bone implant's surface to improve osseointegration. Sandblasting, etching, or machining can create such surface roughness. For titanium bone implants, titanium oxide (TiO_2) nanoparticles are frequently added to titanium to create surface roughness. Osteoblasts (bone cells) adhere to such roughened surfaces quite well, proliferate well, and even deposit calcium phosphates (apatites) (Fig. 6.7). Unfortunately, such behavior is not easily recapitulated in vivo, especially in the long term. More sophisticated patterning of micrometer- or even nanometer-sized structures have been evaluated and showed some improvements.

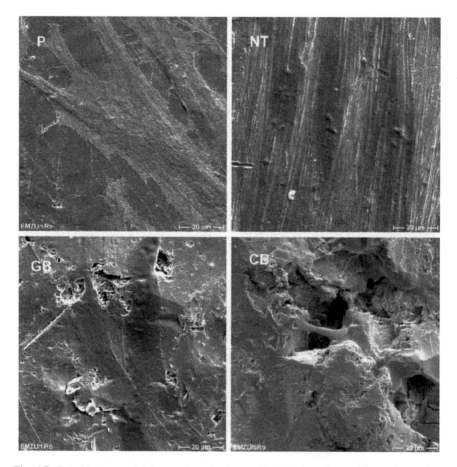

Fig. 6.7 Osteoblasts are added to various titanium oxide (TiO_2) surfaces with varying surface roughness. Root mean square roughness = 0.19 μm (P), 0.54 μm (NT), 1.22 μm (GB), and 6.07 μm (CB). The highest surface roughness (6.07 μm) resulted in ideal osteoblast adhesion, spanning through the ridges, while the osteoblasts on other surfaces are essentially flat. (Lüthen et al., 2005. Reprinted with permission. (C) 2005 Elsevier)

Question 6.1 Increasing the surface roughness on metal implants improves tissue ingrowth. Why?

While some successes have been demonstrated for introducing osteoblasts on metal surfaces (and subsequently ceramics and polymers) toward successful tissue-engineered bone grafts, better material is preferred other than metals, ceramics, or polymers.

To address this issue, many researchers are developing tissue-engineered bone grafts. A 3D biomaterial is preferred to fully recapitulate the 3D structure of bone, where osteoblasts can adhere to, proliferate on, and deposit minerals (calcium phosphates). Naturally, calcium phosphates and other apatites (e.g., calcium carbonates) interact strongly and specifically with osteoblasts and existing bones. Such 3D calcium phosphate scaffold is currently available commercially, for example, from BD, which has been utilized to analyze bone metabolism in vitro and in vivo.

Another famous example is *Bioglass*®, a mixture of silicon, calcium, and sodium oxide at 45%:25%:25% (the rest is phosphorous oxide) (Fig. 6.8). In warm water (and in the human body), the sodium dissolves at the surface, and the rest turns into tiny crystals of hydroxy carbonate apatite (HCA). This material is similar to hydroxyapatite (or hydroxy calcium phosphate), where phosphate is replaced with carbonate. Recall that calcium carbonates are found in clams and corrals. The dissolved sodium and the crystal formation generate porous structure, especially on its surface, which is a preferred surface for osteoblasts to adhere to and proliferate. No complicated microfabrication is necessary as the surface pores are generated spontaneously in warm water. However, it is mechanically weak and easy to be fractured compared to real bones because it is porous and not cross-linked to improve its mechanical strength and toughness.

Fig. 6.8 Bioglass. (Poologasundarampillai et al., 2016. Reprinted with permission. (C) 2016 The American Ceramic Society and John Wiley & Sons)

Bioglass was developed by Larry Hench and had a fascinating history of development. Dr. Hench was on a bus ride from Florida to New York to attend a conference in 1967. On the bus, he met Colonel Klinker, who had just returned from Vietnam War. Colonel Klinker had witnessed numerous amputations (removal of arms and legs) and found that the body rejected many metal and polymer implants. His challenge was to develop a material to help regenerate bone. Dr. Hench's answer to this challenge was Bioglass, and the first result was published in 1971.

6.8   Hydrogels

Hydrogels are one of the most popularly used synthetic scaffolds for tissue engineering applications. As explained briefly in Sect. 6.4, hydrogels are a cross-linked polymer network filled with water. And the polymers are either polysaccharides (alginate, dextran, cellulose, starch, chitosan, and hyaluronan) or proteins (collagen, elastin, etc.). These hydrogels have successfully been utilized as synthetic scaffolds for tissue engineering applications. Despite a substantial amount of research, these natural hydrogels suffer from various problems, mostly due to their inherent differences from mammalian ECM structures. *Nanofibrillar hydrogels* are gaining popularity in recent years, essentially polypeptides that are self-assembled to form fibrillar structures in a few tens of nanometer scale. We can fine-tune molecular self-assembly to create a specific shape. For example, we can change the peptide sequence to precisely control their structures, mechanical properties, etc. We can also incorporate growth factors, cell receptors, enzymes, etc., at the desired location. Synthetic networks can also be cross-linked to improve their mechanical strength. Such design schematics are illustrated in Fig. 6.9.

6.9   Biodegradable Scaffolds

Some biomaterial scaffolds are deliberately designed to be biodegraded. Such scaffold maintains its shape and structure during in vitro culture. Once they are transplanted in vivo, they slowly degrade in the body. Such *biodegradation* may be useful to release growth factors and other bioactive compounds from the scaffold. Also, biodegradation may allow the host body to develop ECM to replace the scaffold newly. Such biodegradation addresses both concerns listed above – matching the exact components of natural ECM and eliminating the nearby ECM interference.

Polylactic acid (PLA), polyglycolic acid (PGA), and polylactic-co-glycolic acid (PLGA) are the most popular biodegradable materials. Their biocompatibility and

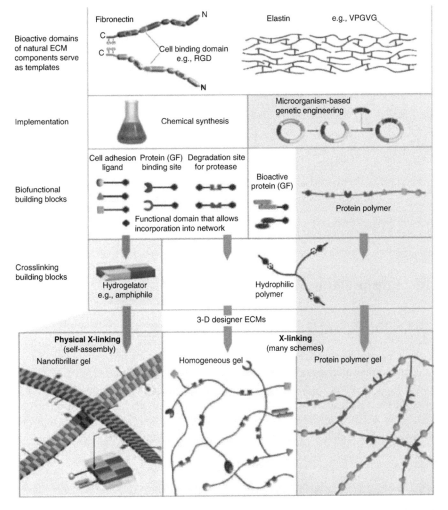

Fig. 6.9 Design strategies for the creation of synthetic biomolecular materials that mimic the complexity of natural ECMs. (Lutolf & Hubbell, 2005. Reprinted with permission. (C) 2005 Springer Nature)

biodegradability have been well-documented. Likewise, many countries (e.g., US FDA) have allowed their use for in vivo clinical applications. Polyhydroxyalkanoates (PHAs), including poly-β-hydroxybutyrate (PHB), have also been studied as biodegradable polymer materials. Bacteria or yeasts usually synthesize PHAs. Specifically for bone, another polymer has also been studied and tested – polypropylene fumarate (PPF). Structures of these biodegradable polymers are shown in Fig. 6.10. You can notice that all of them are polyesters.

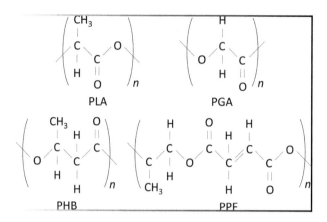

Fig. 6.10 Biodegradable polymers. PLA, PGA, PHB, and PPF

6.10   Encapsulation Scaffolds

Hollow spheres made from the semipermeable membrane can also be used as tissue-engineered scaffolds. Cells are loaded inside such spheres, whose procedure is known as *encapsulation*. Cells are protected from the host's immune cells' attacks as the membrane pores are much smaller than the immune cells. As it primarily protects the transplant from the host's immune attacks, it is also known as *immuno-isolation*, described in Sect. 1.4. Oxygen, nutrients, growth factors, and other cell signaling molecules are free to move in and out of the semipermeable membrane, allowing cells to metabolize and proliferate in a usual manner (Fig. 6.11).

6.11   Laboratory Task 1: Preparation of Various Biomaterial Surfaces

In this task, various biomaterial surfaces are prepared, which can be used for tissue engineering applications.

Objective 1. Preparation of PMMA Surface
PMMA (polymethyl methacrylate) is a popular polymer material that has been used for various applications. PMMA is sometimes preferred for tissue engineering applications over other hydrophobic polymers such as PE, PTFE (and its variant Teflon), and PDMS (which is discussed in the next objective). Hydrophobicity is

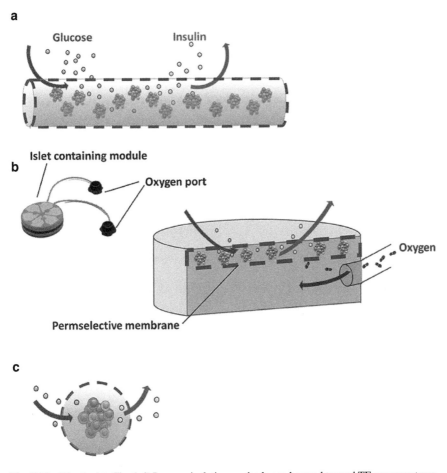

Fig. 6.11 (Identical to Fig. 1.6) Immunoisolation methods can be used toward TE pancreas transplant. Polymeric approaches to reduce tissue responses against devices applied for islet-cell encapsulation. (Hu & de Vos, 2019. (C) 2019 Hu and de Vos. Open access article distributed under the terms of the Creative Commons Attribution 4.0 International License)

typically measured by the water contact angle, which is demonstrated in the next section. The water contact angles of PE, PTFE, and PDMS are around 110° or even higher, indicating strong hydrophobicity. Meanwhile, PMMA's water contact angle is approximately 75°, which is not very different from those of dried protein films – around 40°–60°. PMMA is still easy to make into a particular shape for specific tissue engineering applications and create surface micro- and nanometer patterns.

In this objective, PMMA will be spin-coated on a microscopic coverslip, rather than creating a whole biomaterial from PMMA. *Spin coating* is a popular process for creating a thin film, where a viscous polymer solution is applied on a substrate

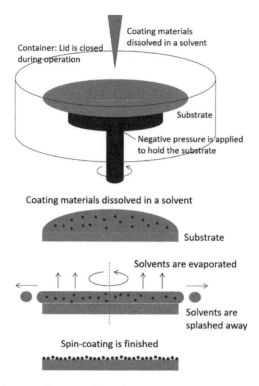

Fig. 6.12 Top: a spin coater; bottom: spin coating process

and spun at a few thousands of RPM (rounds per minutes) for a minute or two. The viscous polymer solution is spread evenly on a substrate, and any excess polymers are dispensed outside the substrate (Fig. 6.12). A vacuum is typically applied to the stage to hold the substrate on a stage during spinning.

Such a thin film is then exposed to UV light with a mask on it. The patterns of the mask are transferred to the thin film. Fragmented films are rinsed away to generate surface micro- or nanometer patterns. This procedure is known as *photolithography* and is still used to create line patterns, posts, wells, etc., on biomaterial surfaces (Fig. 6.13).

In this objective, we will conduct spin coating to create PMMA film on a glass substrate without creating micrometer patterns via photolithography.

1. Set the hot plate to 180 °C.
2. Wipe a microscope coverslip with ethanol and KimWipes. Gloves must be worn at all times to prevent fingerprinting the surfaces.

Fig. 6.13 Photolithography. (Yoon, 2016. Reprinted with permission. (C) 2016 Springer)

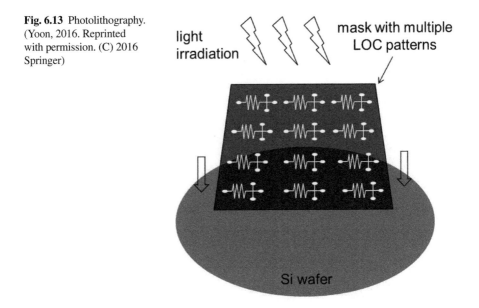

3. Place the coverslip to the center of the stage inside a spin coater.
4. Dispense 30 µL of PMMA solution (4:6 mixture of PMMA and C4 thinner) onto the center of the coverslip using a pipette.
5. Close the lid of the spin coater. Run spin coating at 2000 RPM (round per minute) for 30 s.
6. Take the PMMA-coated glass coverslip from the spin coater. Bake on a hot plate for 1 min.

Objective 2. Preparation of PDMS Surface

PDMS (polydimethylsiloxane) is probably the most widely used material for microfluidic devices, and accordingly, tissue-engineered organ mimics. PDMS is the most popularly used material for soft lithography, where a "mold" is fabricated using conventional photolithography. A PDMS gel is poured onto it, cured and cross-linked, and peeled off from the mold to make replica copies of it (Fig. 6.14).

Once a single mold is made from spin coating and photolithography, you can create multiple replicas in a conventional laboratory without the need for any equipment. Like photolithography, various micro- or nanostructures can be "patterned" on the PDMS surfaces. We can utilize such patterns to induce cellular adhesion, morphogenesis, differentiation, and proliferation, as addressed in Sects. 6.2 and 6.3 (Fig. 6.15).

Fig. 6.14 Left: mixing of silicone base and curing agent; right: vacuuming the substrate. (Yoon JY. 2016. Introduction to Biosensors, 2nd edition. Springer: New York, Chapter 14. https://doi. org/10.1007/978-3-319-27413-3_14. Reprinted with permission. (C) 2016 Springer)

Fig. 6.15 Isolation of collagen type I from a rat tail and coating on a TCP for cell culture

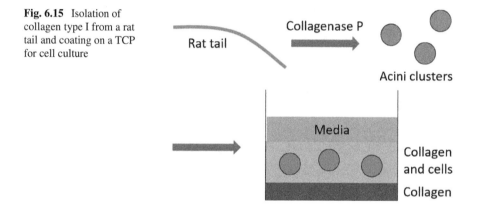

7. Wipe the inside surface of the petri dish with ethanol and KimWipes. Gloves must be worn at all times to prevent fingerprinting the surfaces.
8. In a plastic cup, mix 6 mL silicone base (e.g., Sylgard 184 silicone elastomer base = PDMS) with 0.6 mL curing agent (e.g., Sylgard 184 silicone elastomer curing agent).
9. Pour the mixture on the petri dish. Place the petri dish in a desiccator that is connected to a vacuum pump. Apply vacuum for 20 min. (Normally, a silicon chip with an appropriate surface micro-pattern is placed on the petri dish, and the mixture is poured on top of it. The surface micro-patterns will be transferred to the silicone elastomer, that is, PDMS, and this procedure is known as replica molding, one of the most popular methods of soft lithography.)
10. Place the petri dish with PDMS into the oven at 65 °C and bake for 1 hour.

Objective 3. Smear Coating of Rat Tail Collagen Type I
Modification of biomaterial surface with collagens, specifically rat tail collagen type I, is a popular method to provide an ECM-like and adhesion-friendly surface for mammalian cells, as we have learned from Sect. 6.4 (Fig. 6.15).

11. Prepare a microscope glass slide and clean the surface with ethanol and KimWipes. Gloves must be worn at all times to prevent fingerprinting the surfaces.
12. Dispense 10 μL of collagen solution using a pipette.
13. Use the same pipette tip to smear the collagen droplet until it is completely spread out and smooth on the glass slide surface.
14. Let it dry for 10 min.

6.12   Laboratory Task 2: Contact Angle Measurements

Hydrophobicity, that is, how much hydrophobic the surface is, is essential in assessing biomaterial surfaces. A hydrophobic surface will resist water molecules' adhesion, known as a nonstick surface – an excellent commercial example is Teflon coating for cookware. As mammalian cells and tissues' major component is water, a hydrophobic surface is generally considered not favorable for cell culture and tissue engineering applications. A cell membrane's surface is positively charged from its phospholipid bilayers, and cell adhesion is preferred on a hydrophilic surface. Many proteins are not fully soluble in water, rendering them "relatively" hydrophobic, and they tend to adsorb to hydrophobic surfaces via hydrophobic interactions. Once such hydrophobic interaction-induced adsorption occurs, proteins tend to change their conformations. The inner hydrophobic cores are exposed through unfolding and make even stronger adhesion to the hydrophobic surface, leading to flattening or spreading on the surface (Yoon et al., 1998). This process makes the proteins even more challenging to be removed from the surface, leading to permanent fouling, and often signaling blood coagulation cascade. However, hydrophobic surfaces are favored for some applications, especially for blood vessels and cardiovascular applications, where water resistance (nonstick feature) is preferred.

Hydrophobicity is easily measured by monitoring water contact angle, as depicted in Fig. 6.16. As "hydrophobicity" is always a relative term, the surface of 30° contact angle can be considered hydrophobic if all other materials have contact angles lower than 20°. Nonetheless, the surfaces with 0°–40° contact angles can generally be regarded as hydrophilic, >85° hydrophobic, and 41°–84° intermediate. Glass, metals, and PEG (polyethylene glycol) are hydrophilic; PMMA, PC (polycarbonate), and most protein films are intermediate; PE, PTFE, Teflon, and PDMS are hydrophobic.

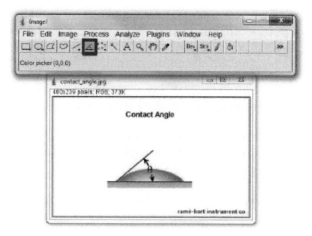

Fig. 6.16 Contact angle measurement in ImageJ

Water contact angles can be measured by a sophisticated contact angle analyzer, which consists of (1) a droplet dispensing system (with a motorized syringe), (2) a positioning stage, (3) a digital camera, and (4) software to analyze the image to obtain water contact angle. As its detection principle is straightforward, you can dispense a water droplet (typically around 10 μL) on any given surface, take a digital image from the side with a smartphone, and determine the contact angle using appropriate software. We have already learned how to download and use the free software ImageJ (from US NIH) to process the microscopic images. ImageJ can also analyze the contact angles from the water droplets' photos, which we will learn in this task.

15. Apply 10 μL water droplets on various biomaterial surfaces using a pipette. If a droplet is challenging to be dispensed on the surface, increase the volume to 20 μL.
16. Take the images of water droplets on various biomaterial surfaces. Open these images in ImageJ.
17. Select the Angle Tool (Fig. 6.16).
18. Draw an angle with the first side along the biomaterial surface and the second side tangent to the droplet edge, as shown in Fig. 6.16.
19. Choose Analyze -> Measure. A new window will pop up showing the list of angle measurements (Fig. 6.17).
20. Compare the contact angles of various biomaterial surfaces. PMMA should be around 75°, PDMS around 110°, and collagen film around 30°.

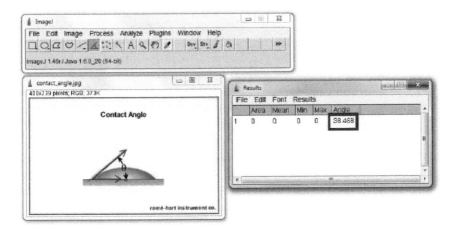

Fig. 6.17 ImageJ determines the contact angle

Question 6.2 What property is measured from contact angle analysis? Why is it important for tissue engineering?

6.13   Laboratory Task 3: Surface Roughness Measurements

The surface roughness of biomaterial surfaces is also crucial in characterizing biomaterial surfaces. For some applications, for example, bone tissue engineering, roughness is mandatory to induce tissue ingrowth (Sect. 6.7). Suppose the biomaterial surfaces are patterned with micro- and nanometer structures. In that case, it is also essential to evaluate the overall roughness of such patterns whether it fits the dimension of a mammalian cell (Sect. 6.2). In this task, we will measure surface roughness from the microscopic images of various biomaterial surfaces.

21. Collect the microscopic images using a benchtop microscope. There is no need to use fluorescence.
22. Download the roughness calculation plugin for ImageJ from https://imagej.nih. gov/ij/plugins/download/Roughness_calculation.class.
23. Install the plugin from the ImageJ menu Plugin -> Install -> choose the downloaded roughness calculation plugin -> restart ImageJ.
24. Open the microscopic image of a biomaterial surface.
25. Select the square tool. Draw a square to cover the most images of biomaterial surface (Figs. 6.18, 6.19 and 6.20).
26. Plugin -> Roughness calculation. A window will pop up, showing the root mean square (RMS) roughness value. Note that the numbers will vary by the magnification. You should adjust the value by using an appropriate length scale.

Fig. 6.18 Smartphone images of water droplets on PMMA (left) and collagen coating (right)

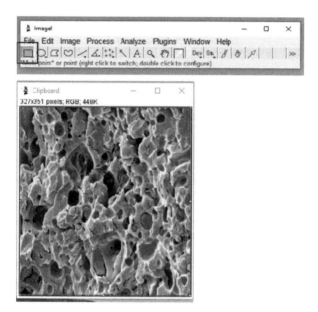

Fig. 6.19 Choosing the area of interest from a microscope image

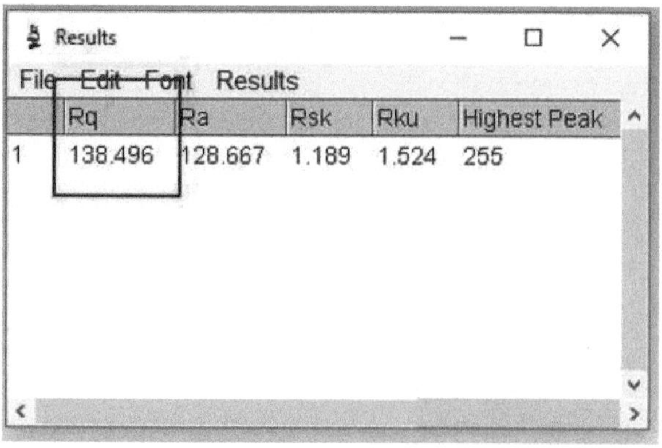

Fig. 6.20 Evaluated RMS roughness from ImageJ plugin

27. Repeat for the other biomaterial surfaces and compare them.

ImageJ plugin provides various other statistical parameters. They are:

– Ra = arithmetical mean deviation.
– Rq = root mean square deviation (this is the one most commonly used).
– Rku = kurtosis of the assessed profile.
– Rsk = skewness of the assessed profile.
– Rv = lowest valley, given by the minimum measurements.
– Rt = total height of the profile.

Review Questions

1. Discuss how 2D versus 3D and length dimensions affect cellular adhesion on biomaterial surfaces and scaffolds.
2. Describe the strengths and weaknesses of synthetic scaffolds over natural scaffolds.
3. Compare various polysaccharides for tissue engineering applications. Why are chitosan and hyaluronan (HA) more popular than the others?
4. Describe and compare different interactions between cells and biomaterial surfaces.
5. Increasing the surface roughness on metal implants improves tissue ingrowth. Why?
6. What is Bioglass? Why is it suitable for bone implants?
7. Compare natural hydrogels and synthetic nanofibrillar hydrogels for their tissue engineering applications.
8. Describe the strengths and weaknesses of biodegradable scaffolds (e.g., PLGA).
9. Graphically illustrate the spin coating process. How can spin coating be used toward improving the biocompatibility of biomaterial scaffold?
10. Graphically illustrate how you can evaluate the contact angles of biomaterial surfaces. What information can be obtained through contact angle measurement for biomaterial scaffold?
11. What property is measured from contact angle analysis? Why is it important for tissue engineering?

References

Hu, S., & de Vos, P. (2019). Polymeric approaches to reduce tissue responses against devices applied for islet-cell encapsulation. *Frontiers in Bioengineering and Biotechnology, 7*, 134. https://doi.org/10.3389/fbioe.2019.00134
Lüthen, F., Lange, R., Becker, P., Rychly, J., Barbara, B. U., & Nebe, J. G. (2005). The influence of surface roughness of titanium on β1- and β3-integrin adhesion and the organization of fibronectin in human osteoblastic cells. *Biomaterials, 26*, 2423–2440. https://doi.org/10.1016/j.biomaterials.2004.07.054
Lutolf, M. P., & Hubbell, J. A. (2005). Synthetic biomaterials as instructive extracellular microenvironments for morphogenesis in tissue engineering. *Nature Biotechnology, 23*, 47–55. https://doi.org/10.1038/nbt1055

Poologasundarampillai, G., Lee, P. D., Lam, C., Kourkouta, A. M., & Jones, J. R. (2016). Compressive strength of bioactive sol-gel glass foam scaffolds. *International Journal of Applied Glass Science, 7*, 229–237. https://doi.org/10.1111/ijag.12211

Yoon, J.-Y. (2016). Chapter 14. In *Introduction to biosensors* (2nd ed.). New York. https://doi.org/10.1007/978-3-319-27413-3_14

Yoon, J.-Y., Kim, J.-H., & Kim, W.-S. (1998). Interpretation of protein adsorption phenomena onto functional microspheres. *Colloids and Surfaces B: Biointerfaces, 12*, 15–22. https://doi.org/10.1016/S0927-7765(98)00045-9

Chapter 7
Focal Adhesion

In the previous chapter, we learned about various biomaterial surfaces used for tissue engineering applications. Successful biomaterial surfaces should accommodate cellular adhesion and proliferation, which occur naturally on the extracellular matrix (ECM). Therefore, biomaterial surfaces that mimic ECM have been preferred for tissue engineering applications and investigated extensively. Adhesion of cells on a biomaterial surface can occur in a nonspecific or a specific manner. Of course, specific adhesion is preferred, and the most crucial cellular adhesion on biomaterial surfaces (and on ECM) is focal adhesion. In this chapter, we will specifically learn this focal adhesion, especially on various biomaterial surfaces.

Inquiry 1. Can you explain why specific cellular adhesion is better than nonspecific adhesion?

Inquiry 2. Can you list the names of proteins involved in focal adhesion?

7.1   What Is Focal Adhesion?

In Chap. 2, we have briefly learned focal adhesion. *Focal adhesion* is an important, specific cellular adhesion mechanism for anchorage-dependent mammalian cells. Focal adhesion occurs naturally on extracellular matrix (ECM), and we want the same focal adhesion on biomaterial surfaces.

Figure 7.1 shows the schematic diagram of focal adhesion, briefly explained in Sect. 2.3 of Chap. 2. The top portion represents the inside of a mammalian cell, and the bottom part represents the biomaterial surface. To encourage better cellular adhesion, you will need to modify the biomaterial surface with an ECM protein. In mammalian tissues, the bottom portion is, of course, simply ECM.

A membrane protein, integrin, mediates focal adhesion. It has two similarly structured subunits floating on the cell membrane (phospholipid bilayer) (Fig. 7.2).

Integrin binds to several different ECM proteins, including:

© Springer Nature Switzerland AG 2022

J.-Y. Yoon, *Tissue Engineering*, https://doi.org/10.1007/978-3-030-83696-2_7

Fig. 7.1 (Identical to Fig. 2.4) Focal adhesion

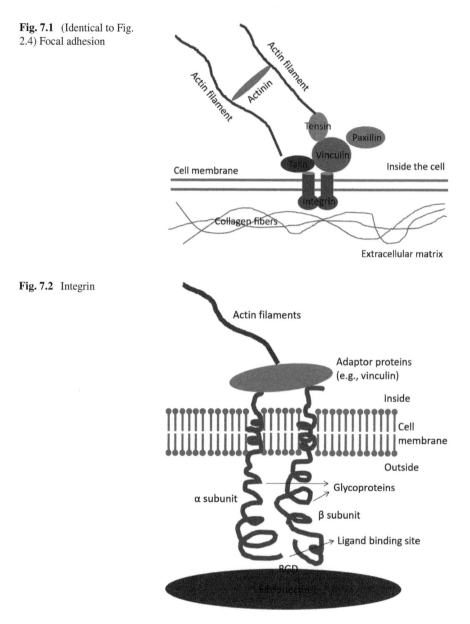

Fig. 7.2 Integrin

- *Collagen*: the major fibrous backbone of ECM.
- *Fibronectin*: cell–ECM glue protein.
- *Vitronectin*: like fibronectin but smaller in size.
- *Laminin*: a major component of the basement membrane (BM) – one type of ECM.

Integrin binds specifically to the tripeptide sequence *RGD* (*arginine–glycine–aspartic acid*), which can be found in all the above ECM proteins.

Many biomaterial surfaces have been coated with these ECM proteins and are sometimes made entirely with these proteins. A good example is *Matrigel* from BD, made from cell culture. It is rich in laminin and collagen and a good approximation of the basement membrane. A *basement membrane* (*BM*) is a thin sheet-shaped ECM, and one type of ECM. As many biomaterial surfaces are flat and sheet-like, the basement membrane approximation is a good strategy in fabricating a biomaterial surface.

Alternatively, RGD-containing oligopeptides can also be used, for example, GRGDSP (glycine–arginine–glycine–aspartic acid–serine–proline) or GRGDSPK (glycine arginine–glycine–aspartic acid–serine–proline–lysine). The last part, lysine (K), can be added or omitted depending on the biomaterial surface's charge, for example, a negatively charged biomaterial surface can easily accommodate the positively charged lysine (K).

Question 7.1 Integrin can bind to (choovse one):

1. Actin filaments	Yes	No
2. Cadherin	Yes	No
3. Collagen	Yes	No
4. Fibronectin	Yes	No
5. Laminin	Yes	No

Once integrin–ECM protein binding (specifically RGD motif) occurs, integrin changes its conformation and triggers its binding to adaptor proteins such as *actinin*, *talin*, and *vinculin*, inside the cells. These adaptor proteins then bind to actin filaments (Fig. 7.1). Vinculin is always found where focal adhesion occurs. Therefore, staining vinculin with anti-vinculin-FITC or anti-vinculin-TRITC is common in locating the focal adhesion points within cells (Fig. 7.3).

Therefore, a strong connection from ECM proteins – integrins – adaptor proteins (including vinculin) – to actin filaments is formed in focal adhesion. You should note that focal adhesion connects the biomaterial surface (ECM proteins) to the cell membrane and the actin filaments. In contrast, nonspecific adhesion involves the connection between the biomaterial surface and cell membrane but not to the actin filaments (Fig. 7.4).

7.2   Cell Adhesion and Proliferation on Biomaterial Surfaces

Figure 7.5 schematically illustrates the cell–surface and cell–cell interactions. The cell shown below makes a focal adhesion with the laminin-coated biomaterial surface, where laminin is the significant component of the basement membrane (thin sheet-type ECM). Laminin is bound to the integrins in the cell membrane, and through adaptor proteins (especially vinculin), they are connected to the actin filaments (cytoskeleton). Fibronectin, a cell–ECM glue protein, is also bound to the integrins, again via focal adhesion, that is, connected to the actin filaments. Collagens, major backbone fibrous proteins in ECM, are also attached to the

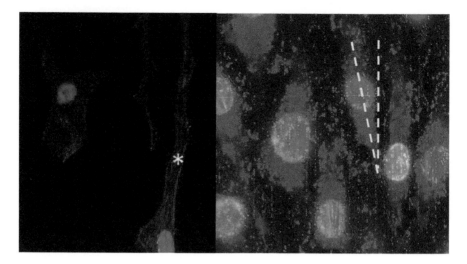

Fig. 7.3 Left: mouse fibroblasts make focal adhesions on a wave micrometer pattern. Nuclei are stained with DAPI (blue), actin filaments with phalloidin-TRITC (red), and vinculin with anti-vinculin and anti-IgG-FITC (green). (Gamboa et al., 2013. Reprinted with permission. (C) 2013 Elsevier). Right: human umbilical vein endothelial cells (HUVECs) make focal adhesions on a nanoparticle-nanowell composite surface. Staining methods are identical to the left. (Tran et al., 2013. Reprinted with permission. (C) 2013 John Wiley and Sons)

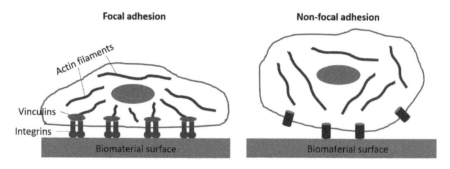

Fig. 7.4 Good adhesion strength is demonstrated in focal adhesion on the biomaterial surface (left), where the surface is linked to actin filaments (cytoskeleton). Meanwhile, insufficient adhesion strength is demonstrated in nonfocal (nonspecific) adhesion, where the surface is connected only to the cell membrane but not to the actin filaments. The presence of vinculin can confirm successful focal adhesion

integrins via focal adhesion, that is, connected to the actin filaments. In this case, collagens are not coated on the biomaterial surface, but they connect two cells to form a multilayer of cells on the biomaterial surface.

Similarly, *cadherins* are connecting two cells directly, which is not focal adhesion. The addition of collagens to the cells (not on the surface) and cadherin-mediated cell–cell adhesion allow the formation of a multilayer of cells on the biomaterial surface, which may be useful for specific tissue engineering

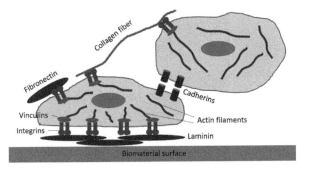

Fig. 7.5 Overview of cell–surface and cell–cell interactions

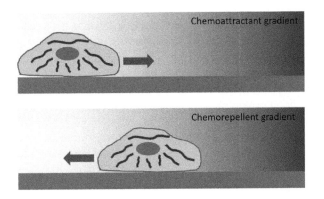

Fig. 7.6 Chemoattractants, chemorepellents, and chemotaxis

applications. Once cells properly adhere, they exhibit spread (not rounded) morphology, leading to optimum metabolism. In turn, this will trigger the production of growth factors, allowing the cells to proliferate on the biomaterial surface.

7.3   Cell Migration After Focal Adhesion

Once cells make successful focal adhesion, they exhibit spread (not rounded) morphology and start proliferating. Before that, cells must be migrated over the surface to reach the confluency on biomaterial surfaces. Cell migration is a cyclic process in which a cell extends protrusions at its front and retracts its trailing end. Cells are migrating toward or against the gradient of chemicals (including proteins). If cells migrate toward such chemicals (including proteins), they are called *chemoattractants*. If they migrate away from them, they are called *chemorepellents* (Fig. 7.6). And these processes are called *chemotaxis*.

Chemotaxis should not be confused with molecular diffusion. It is an active transport mechanism mediated by chemotaxis receptors on cell membranes.

Chemotaxis receptor binds to a portion of a chemoattractant, called chemoattractant moiety. Chemorepellents are substances expressing adverse migratory effects.

Here is a list of chemoattractants:

- *Formyl peptides*: These are very short peptides (di-, tri-, tetra-, etc.) that originated from bacteria. They can be either released from bacteria or are remnants after decomposition. In the human body, bacteria are generally not wanted. The white blood cells (neutrophils, monocytes, etc.) are attracted to them through chemotaxis to neutralize them (inflammatory response). Note that monocytes escape from the blood vessel and turn into macrophages in the tissue.
- *Complementary 3a (C3a)* and *complementary 5a (C5a)*: These proteins are parts of the complement cascade. *Complementary cascade* is a complementary system to the human body's normal immune response and is often considered a part of the immune system. Like formyl peptides, C3a and C5a act as chemoattractants to the white blood cells (neutrophils, monocytes, etc.).
- *Chemokines*: These are the cytokines that can function as chemoattractants or chemorepellents. *Cytokine* is a broad category of small proteins produced from cells that have an essential role in cell signaling. CXC chemokines and CC chemokines are good examples that can attract neutrophils (CXC chemokines) and monocytes (CC chemokines).
- *Leukotrienes*: These are the inflammatory mediators released by white blood cells (= *leukocytes*). The most famous example is leukotriene B4 (LTB4). It elicits cellular adhesion, chemotaxis, and leukocytes' (white blood cells') aggregation, especially during inflammation and allergy response.

Cell migration from left to right shown in Fig. 7.7 occurs in the following order:

1. *Polarization*: Migration-promoting agents or chemoattractants (or chemorepellents) are found on the right side that induces an initial polarization. This process creates *lamellipodium* (lamelli = thin sheet; podium = foot; the plural form is lamellipodia). As its name indicates, the initial protrusion is very thin.
2. *Protrusion*: Actins are polymerized in the right side of a cell, generating actin filaments. This process will create a pronounced protrusion of a cell to the right.
3. *Traction*: Focal adhesions are formed on the newly created protrusion area.

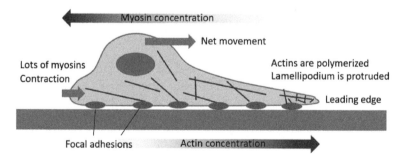

Fig. 7.7 Cell migration on biomaterial surface

Fig. 7.8 Lamellipodia of
bone cells on collagen
fiber-coated surface.
(Friedrichs et al., 2007.
Reprinted with permission.
(C) 2007 Elsevier)

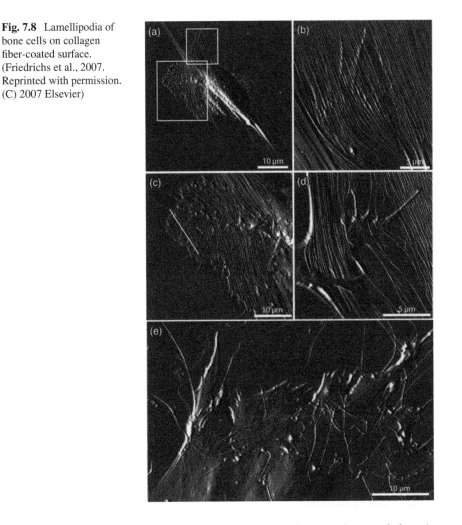

4. *Retraction*: On the other side (left) of a cell, *myosin* induces the cytoskeleton's
   contraction.

Figure 7.8 shows an example of cell migration (human bone cells) on a surface
coated with collagen. Images are taken by atomic force microscopy (AFM). The squared
areas in (a) indicate lamellipodia that resemble foot but in quite thin sheets. (b) and (c)
are the images at higher magnification. Smaller extrusions can also be found, called
filopodia (filo = thread; singular = filopodium). Both lamellipodia and filopodia extend
underneath the collagen fibers coated on the surface as shown in (d) and (e).

The formation of lamellipodia and filopodia is also graphically summarized in
Fig. 7.9. Actin filaments form loose meshwork in the lamellipodium. The filopodium is characterized by a smaller bump, typically in radial bundles. In the other
parts of a cell, actin fibers are well polymerized and organized, forming *stress fibers*.
Focal adhesion sites are indicated in green. Not many focal adhesions are found in

Fig. 7.9 Polarized
fibroblast on a surface,
showing lamellipodium
and filopodium

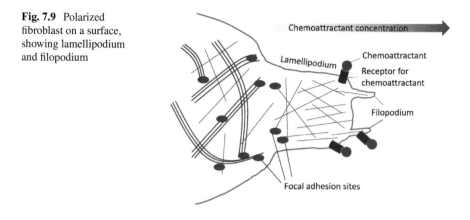

lamellipodia and filopodia. Once sufficient focal adhesions and actin polymerization are made, this region is no longer called lamellipodia. Actin filaments are polarized to the right, with a fast polymerizing end.

7.4   Morphogenesis

Once cells are adhered and migrated, they should go through morphogenesis.
Morphogenesis is essential to assign cells into specific roles (cell types are not
changed with somatic cells, while they are with stem cells). They can be guided by
the structural cues from ECM or chemical signals from neighboring cells.
Specifically:

– *Diffusion*: One cell produces a morphogenetic inducer and diffuses to the other
  cell, inducing a change in the other cell.
– *ECM*: ECM (and subsequently biomaterial surface) can induce morphogenesis
  directly or indirectly.
– *Cell–cell contact*: One cell induces morphogenesis to the other cell in direct
  contact.

Morphogenesis is a complicated process. Cells and ECM are affecting each other
toward correct morphogenesis. To induce proper and appropriate morphogenesis in
tissue engineering applications, you will need to create a biomaterial surface that
can generate structural cues and mechanical forces to the cells, similar to the natural
ECM. Mechanical forces include stretching, compression, shear, and even voltage
differentials. Cells have "sensors" that can respond to such cues and stimuli. These
factors are considered *epigenetic*, that is, outside the genome. Besides, cells also
react to chemicals (including proteins) toward correct morphogenesis.

7.5   Laboratory Task 1: Fluorescence Staining of Focal Adhesion

In this task, vinculins at focal adhesion sites will be stained with anti-vinculin and FITC. Actin filaments will also be stained with phalloidin-TRITC. Nuclei will optionally be stained with DAPI.

Objective 1. Kill, Fix, and Perforate Cells

1. Anchorage-dependent cells are cultured on 24-well tissue culture plates. Refer to the previous chapters for cell culture.
2. Wear gloves and sterilize with ethanol. Sterilize the biosafety cabinet. Then, transfer the 24-well culture plate into the biosafety cabinet.
3. Rinse each well twice with 250 μL washing buffer (1 × PBS with 0.05% Tween 20; Tween 20 is surfactant).
4. Immediately add 100 μL of 4% paraformaldehyde to kill and fix the cells to the surface. Incubate at room temperature in the biosafety cabinet for 15 min.
5. Remove all paraformaldehyde and rise each well twice with 100 μL washing buffer.
6. Add 100 μL of 0.1% Triton X-100 (another surfactant) to each well and incubate for 5 min at room temperature to perforate the cells' membranes.
7. Remove all Triton X-100 and rise each well twice with washing buffer.

Objective 2. Anti-Vinculin Treatment

In this objective, anti-vinculin will be added to the fixed and treated cells, which will bind to the vinculins (indicating focal adhesion sites). Most anti-vinculins are not sold with fluorescent dye stained. We will additionally stain a fluorescent dye in Objective 3.

8. Dilute the anti-vinculin solution in a blocking buffer (1× BPS with 1% BSA) at 2:250, for example, 12 μL of anti-vinculin and 1488 μL blocking buffer in a 2 mL centrifuge tube. Add 250 μL of the anti-vinculin solution to each well and incubate at room temperature for 1 h.
9. Remove anti-vinculin and rinse each well twice with 250 μL washing buffer.

Objective 3. Anti-IgG-FITC and Phalloidin-TRITC

Anti-vinculin will be stained with FITC using anti-IgG. While there are several different forms of antibodies, IgG is the most common type (accounting for ~80% of antibodies), and most commercial antibodies are IgG's. Therefore, anti-IgG will bind to anti-vinculin, and you can easily purchase anti-IgG that are tagged with FITC. In this task, you will also stain actin filaments with phalloidin-TRITC (refer to Chap. 4) to ensure that vinculins can be found only at the actin filaments.

10. Cover the entire plate in aluminum foil during all incubations in this and next objectives (Objectives 3 and 4) to prevent photobleaching of the fluorescent dyes.
11. Dilute the anti-IgG-FITC (FITC-conjugated secondary antibody) solution and phalloidin-TRITC (TRITC-conjugated phalloidin) solution *together* in a blocking buffer at 1:250, for example, 1 μL of the anti-IgG-FITC solution, 1 μL of the phalloidin-TRITC solution, and 248 μL of blocking buffer in a 1 mL of 2 mL centrifuge tube. Add 100 μL of combined dye solution to each well and incubate at room temperature for 30 min.
12. Remove the combined dye solution and rinse each well three times with washing buffer, with 5 min of incubation at room temperature in between each wash cycle.

Objective 4. DAPI (Optional)
Finally, DAPI will be added to stain cell nuclei.

13. Dilute the DAPI (nuclei counterstain) solution in a blocking buffer at 1:250, for example, 1 μL of DAPI solution and 249 μL of blocking buffer in a 1 mL or 2 mL centrifuge tube. Add 100 μL of DAPI solution to each well and incubate at room temperature for 5 min.
14. Remove the DAPI solution and add 380 μL of 1× PBS to each well to prevent the finished cells from drying. If not imaging immediately, seal the entire plate with sealing film (parafilm), cover it in aluminum foil, and place it in a refrigerator, with 5 min of incubation at room temperature in between each wash cycle.

7.6   Laboratory Task 2: Fluorescence Imaging of Focal Adhesion

15. In a dark room, turn on the fluorescence microscope, including the fluorescence light source, the camera, and the computer. To prevent photobleaching, you must close the fluorescence light aperture and block the light path throughout the experiment, except when you capture the image.
16. Place the well plate in a position (Fig. 7.10).
17. Place the FITC (blue excitation and green emission) filter. Examine through the eyepiece. You can observe a small amount of green light, indicating the presence of vinculin, that is, focal adhesion sites. Switch the viewing path of the microscope from the eyepiece to the camera. Capture the image using the camera and store it on a computer. Then, switch the viewing path back to the eyepiece. Close the fluorescence light aperture to prevent photobleaching.
18. Place the TRITC (green excitation and red emission) filter and repeat step 17. You can image actin filaments.

Fig. 7.10 Well plates are placed on a fluorescence microscope with appropriate fluorescence excitations

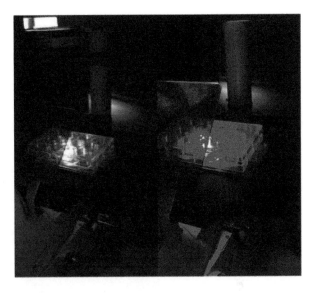

19. (Optional) Place the DAPI (UV excitation and blue emission) filter and repeat step 17. You can image cell nuclei.
20. If necessary, apply pseudo-colors of green to vinculin staining, red to actin filament staining, and blue to nucleus staining using ImageJ. Overlay the images again using ImageJ.

   Two different sets of images are shown in Figs. 7.11 and 7.12. These are the true color images (not with pseudo-color), and the images are not as brilliant as those shown earlier in this chapter. In Fig. 7.11, cells are confluent, and vinculins can only be found where actin filaments exist. As the vinculin image is originally quite dim, brightness and contrast are adjusted, creating a blurry and unclear image. The amount of vinculin is substantially smaller than that of actin filament for the anchored cells. Also, the vinculin staining itself is less efficient (involving two steps of reagent addition – anti-vinculin and anti-IgG-FITC) than the actin filament staining (involving only one step of reagent addition – phalloidin-TRITC). In Fig. 7.12, where cells are not confluent, it is more apparent that vinculins can only be found where actin filaments exist.

Question 7.2 How can you fluorescently stain the following subcellular components of anchorage-dependent mammalian cells? Include the names of fluorescent dyes and receptor molecules. Note that the three fluorescent images will be superimposed together.

1. Nuclei
2. Actin filaments
3. Vinculin

Fig. 7.11 Fluorescence microscope images of anchorage-dependent cells (confluent) on a tissue culture plate. Top left: vinculin-stained; top right: phalloidin-stained (actin filaments); bottom left: DAPI-stained (nuclei); and bottom right: overlayed

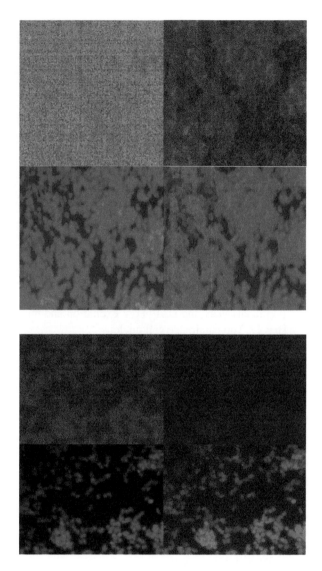

Fig. 7.12 Fluorescence microscope images of anchorage-dependent cells (not confluent) on a tissue culture plate. Top left: vinculin-stained; top right: phalloidin-stained (actin filaments); bottom left: DAPI-stained (nuclei); and bottom right: overlayed

Review Questions

1. What is focal adhesion? What protein is used in identifying focal adhesion?
2. What is the function of fibronectin in ECM? How is it different from laminin?
3. Sketch a diagram of focal adhesion, including ECM protein, integrin, vinculin, and actin filaments.
4. Why is vinculin used as an indicator for focal adhesion?
5. To identify focal adhesion, you should stain both vinculin and actin filaments. Briefly explain why. Graphically illustrate all molecules used for fluorescently staining vinculin and actin filaments.

6. How does RGD-containing oligopeptide improve the scaffold's biocompatibility?
7. Why is focal adhesion stronger than nonspecific adhesion?
8. Sketch a diagram of how (1) collagen, (2) fibronectin, (3) laminin, (4) cadherin, and (5) integrin are involved with cell–cell or cell–ECM binding.
9. What is chemotaxis? How is it different from molecular diffusion?
10. Explain the cell migration process.
11. Identify lamellipodia and filopodia from the image of cell adhesion.

References

Friedrichs, J., Taubenberger, A., Franz, C. M., & Muller, D. J. (2007). Cellular remodelling of individual collagen fibrils visualized by time-lapse AFM. *Journal of Molecular Biology, 372,* 594–607. https://doi.org/10.1016/j.jmb.2007.06.078

Gamboa, J. R., Mohandes, S., Tran, P. L., Slepian, M. J., & Yoon, J.-Y. (2013). Linear fibroblast alignment on sinusoidal wave micropatterns. *Colloids and Surfaces B: Biointerfaces, 104,* 318–325. https://doi.org/10.1016/j.colsurfb.2012.11.035

Tran, P. L., Gamboa, J. R., McCracken, K. E., Riley, M. R., Slepian, M. J., & Yoon, J.-Y. (2013). Nanowell-trapped charged ligand-bearing nanoparticle surfaces: A novel method of enhancing flow-resistant cell adhesion. *Advanced Healthcare Materials, 2,* 1019–1027. https://doi.org/10.1002/adhm.201200250

Chapter 8
Contact Guidance and Cell Patterning

Focal adhesion is an essential mechanism for cells to be anchored on the extracellular matrix (ECM). We can also utilize focal adhesion for biomaterial surfaces toward tissue engineering applications. In natural tissue, however, cells are not randomly assembled on ECM structure – they are aligned to the ECM structure's shape, known as contact guidance. And we can implement this contact guidance on biomaterial surfaces to create a cell–ECM complex that mimics the natural tissue. Biomaterial surfaces can also be patterned in a specific structure to accommodate the cells aligned and assembled in the desired manner through contact guidance. This process is called cell patterning.

Inquiry 1. Can you explain contact guidance in your word?

Inquiry 2. Have you heard about cell patterning? How can it be utilized toward tissue engineering?

8.1 What Is Contact Guidance?

In Chap. 7, we have learned focal adhesion. Focal adhesion is essential for the anchorage-dependent cells to be optimally adhered to the extracellular matrix (ECM) structure, metabolically behave optimally, differentiate into appropriate cell types if necessary, and proliferate to form an optimal tissue. However, to create a fully functioning tissue, cells must be organized in a specific shape and pattern. While there are many factors in regulating such organization, they are primarily dictated by the shape and structure of ECM as cells' anchorage and subsequent assembly are "guided" by the shape of ECM. And this process is called *contact guidance*.

In tissue engineering, we will use biomaterial surfaces in place of ECM. As expected, we can create a specific shape and structure on a biomaterial surface so that the cells can be anchored on such a surface in a controlled and organized

© Springer Nature Switzerland AG 2022
J.-Y. Yoon, *Tissue Engineering*, https://doi.org/10.1007/978-3-030-83696-2_8

manner. This process is a crucial "engineering" aspect of tissue engineering as we can "design" such shape and structure, "fabricate" the tissue, and make them function similarly to the natural tissue.

Figure 8.1 shows a conceptual illustration of contact guidance. In the left, grooves with micrometer or nanometer widths (separated by micrometers or nanometers) are fabricated, where the cells are anchored and aligned to the direction of such grooves. Note that cells are not yet confluent for this case, and they are elongated along the length of grooves. In the right, micrometer-sized wells are fabricated, where the cells are anchored onto each well. Again, cells are not yet confluent and not elongated but rather have rounded morphology. Each groove or well may be coated with ECM proteins such as collagen, "glue" proteins (fibronectin, vitronectin, and laminin), or RGD-containing peptide to promote focal adhesion. However, focal adhesion may still occur without such ECM proteins or RGD-containing peptides, which will be discussed later in this chapter.

8.2   Contact Guidance on Basement Membrane

Contact guidance can frequently be observed on the cells anchored on the *basement membrane* (*BM*), briefly explained in Sect. 6.4. The basement membrane is a thin, membrane-like, sheet-shaped ECM, where the cells are anchored to one side. As the structure is porous (membrane), it allows the transport of cell signaling molecules. While many ECM proteins can be found in the basement membrane, the prime ECM glue protein of basement membrane is laminin, which also has an RGD peptide sequence and can bind to integrin to form focal adhesion. Matrigel® from BD Biosciences is a basement membrane mimic (already explained in Sect. 6.4). It has popularly been used as a surface for growing and anchoring mammalian cells in an in vitro environment. For tissue engineering applications, it is important to recapitulate the component and structure of the basement membrane. Figure 8.2 shows the electron microscopy images of various basement membranes.

Fig. 8.1 Contact guidance on micrometer or nanometer grooves (left) or a micrometer well array (right)

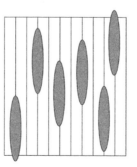 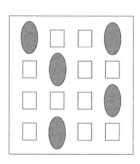

Fig. 8.2 Electron microscope images of the basement membrane of human cornea. (Abrams et al., 2000. Reprinted with permission. (C) 2000 Lippincott Williams & Wilkins)

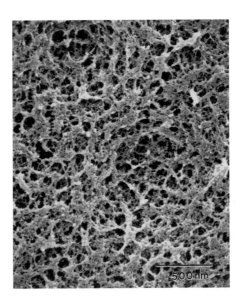

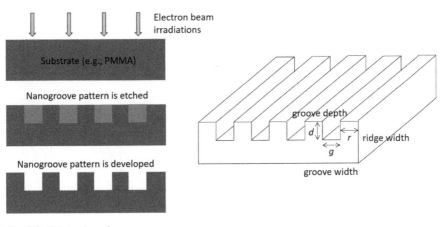

Fig. 8.3 Fabrication of nanogrooves

8.3   Nanogrooves

Nanogrooves are multiplexed parallel grooves typically fabricated on silicon or silicone-based substrates, whose groove widths are in the nanometer scale (Fig. 8.3). While they are not structurally similar to the basement membranes shown in

Fig. 8.2, their structural dimensions are similar to those of basement membranes. Good anchorage, alignment, proliferation, and metabolism can be observed for mammalian cells on these nanogrooves. As explained in Chap. 6, they are typically fabricated using photolithographic techniques.

As explained earlier in Sect. 6.2, each nanogroove's width and the distance between nanogrooves are essential in dictating focal adhesion and their proliferation on them. Figure 8.4 graphically illustrates how mammalian cells are anchored and aligned on two different nanogroove surfaces. Bigger nanogrooves (4000 nm = 4 μm – these are no longer "nano" grooves but "micro" grooves) typically lead to the better proliferation of mammalian cells, whose sizes are usually larger than 10 μm. However, when the nanogroove surface is exposed to shear flow, that is, a flow parallel to the surface, anchored cells can be removed from the surface via shear force more easily with the bigger grooves than with the smaller grooves. Such shear force can easily be found in many tissues, for example, blood vessels, lymphatic vessels, liver sinusoids, kidney tubules, etc.

While it is possible to coat the inner surfaces of nanogrooves with ECM proteins or RGD-containing peptides to promote focal adhesion, mammalian cells do not necessarily need such coatings to accomplish focal adhesion on nanogrooves. Figure 8.5 graphically illustrates how it is possible. An epithelial cell is anchored and spread nicely on a nanogroove surface (i.e., focal adhesion), with the expected alignment toward the direction of the nanogrooves (i.e., contact guidance). They can also create lamellipodia and filopodia. The lateral view of the cell shows that the cell is protruding inside each nanogroove. Under normal focal adhesion, integrin in the cell membrane binds to the RGD motif in the ECM protein, and this binding alters

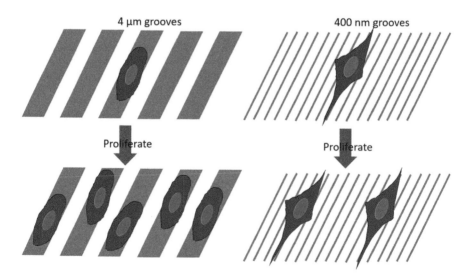

Fig. 8.4 Proliferation and adhesion of mammalian cells on nanogrooves. Better proliferation but poor adhesion is observed on wider nanogrooves, while poor proliferation but better adhesion is observed on narrower nanogrooves

Fig. 8.5 Focal adhesion on the nanogrooves without ECM protein or RGD-containing peptide

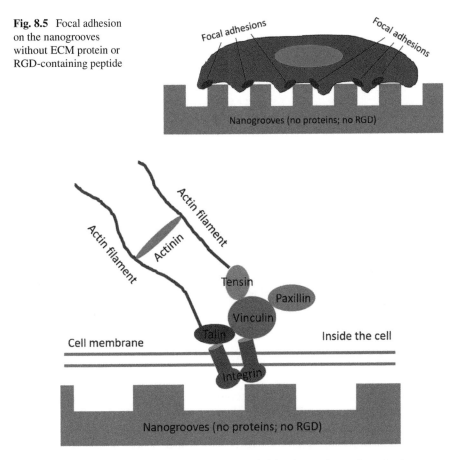

Fig. 8.6 Focal adhesion on a nanogroove. Integrin slightly changes its confirmation in contact with a nanogroove, creating a focal adhesion despite no ECM protein or RGD motif on the surface

the conformation of the integrin. This conformational change triggers adapter proteins' binding inside the cell, most notably vinculin, which connects the integrin to the actin filaments. Hence, the focal adhesion can still be made without integrin-RGD motif binding. This schematic is graphically illustrated in Fig. 8.6. It is also possible to create focal adhesion on other micro- and nanostructures on biomaterial surfaces (as well as to create contact guidance).

Question 8.1 Focal adhesion can form between fibroblasts and nanogrooves without the RGD peptide sequence on them. How can it happen?

Figure 8.7 shows an example of the fibroblasts' focal adhesion and contact guidance on a nanogroove surface. Cells exhibit rounded (not spread) morphology on a smooth surface, and they are not aligned. On a nanogroove surface, however, they exhibit spread and elongated morphology (indirectly indicating that focal adhesions are made), and they are aligned to the direction of the nanogrooves.

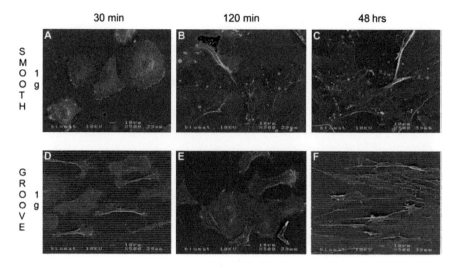

Fig. 8.7 Fibroblasts are cultured on smooth versus nanogroove surfaces, imaged by scanning electron microscope (SEM). (Loesberg et al., 2007. Reprinted with permission. (C) 2007 John Wiley and Sons)

8.4   Shear Flow-Resistant Composite Nanosurfaces

It is difficult to achieve both sufficient proliferation and flow-resistant adhesion on nanostructured surfaces, as discussed in the previous section. Many different strategies have been proposed, and a couple of simple solutions are explained here.

Instead of straight nanogrooves, "wave" nanogrooves can be patterned, as shown in Fig. 8.8. As the dimension of a mammalian cell is much bigger than the size of each wave, cells cannot perfectly fit the shape of wave patterns. Instead, they are aligned straight to the wave pattern, where a cell makes contact outside and inside the groove alternatively. As such, cells can still be anchored very firmly to the surface in a flow-resistant manner, while sufficient confluency can also be demonstrated.

We can also use a composite nanostructured surface. A repeating pattern of nanometer-sized wells has frequently been used to induce focal adhesion and contact guidance. However, their flow resistance has not been proven successful compared to the nanogrooves. Negatively charged nanoparticles can be added to this nanowell pattern and trapped in each nanowell. When mammalian cells are introduced, they endeavor to internalize the nanoparticles but failed to do so due to strong electrostatic attraction between the nanoparticles and the bottom substrate surface. As a result, mammalian cells can remain anchored on the surface even under high shear flow conditions (Fig. 8.9).

Fig. 8.8 Fibroblasts' contact guidance and focal adhesion on wave groove patterns. (Gamboa et al., 2013. Reprinted with permission. (C) 2013 Elsevier)

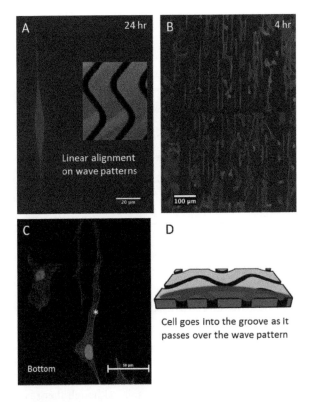

Fig. 8.9 Fibroblasts are anchored and aligned to the nanoparticle–nanowell composite surface. (Tran et al., 2013. Reprinted with permission. (C) 2013 John Wiley and Sons)

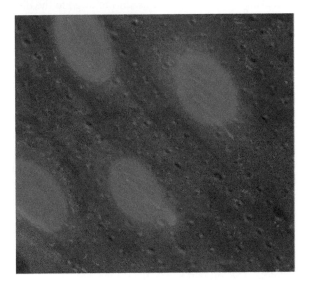

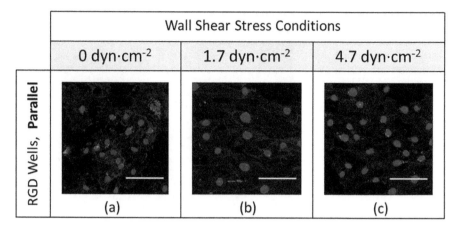

Fig. 8.10 Contact guidance to the direction of flow. (McCracken et al., 2013. (C) 2013 McCracken et al. Open access article distributed under the terms of the Creative Commons Attribution 4.0 International License)

8.5   Contact Guidance by Flow

While cells are aligned to the direction of nanogrooves, they can also be aligned to the direction of flow. In Fig. 8.10, cells are added to the nanowell surface pattern, where each well is coated with RGD-containing peptides. Under no flow, mammalian cells do make focal adhesion, although there is no apparent alignment. With increasing the shear flow, cells start to align to the direction of flow and proliferate better. Cells are more elongated to the direction of flow with the increased shear flow.

8.6   Cell Patterning via Lithographic Protein Patterns

Mammalian cells can be patterned on a biomaterial surface in a controlled manner. This *cell patterning* is somewhat different from contact guidance as the cells are patterned in a more complicated pattern and mostly involve the use of ECM proteins or RGD-containing peptides as "glue." The basic concept of cell patterning is shown in Fig. 8.11, where the glues (ECM proteins or RGD-containing peptides) are printed in an array format. The other area of the surface will be made less cell-friendly, for example, hydrophobic or coated with passivating molecules. When cells are introduced, they will make focal adhesions only on the glue-printed arrays.

Arrays of such ECM proteins or RGD-containing peptides can be printed in a lithographic manner if a high resolution of printing is necessary. Figure 8.12 shows an example. The left part shows the stencil, showing a repeated pattern of 50 μm diameter holes. This stencil is placed on top of a substrate, and collagen solution is added to it. Collagens are printed only through the holes, creating an octagon-shaped array. By changing the stencils, 30 μm, 50 μm, and 100 μm spots of collagens can be printed. Cells are added later to generate cell patterns in the desired form (Fig. 8.12, right).

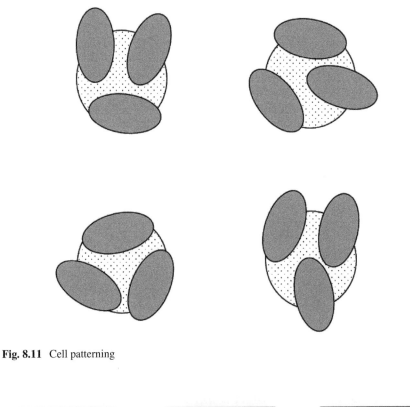

Fig. 8.11 Cell patterning

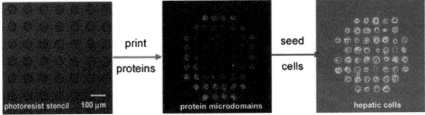

Fig. 8.12 Creation of a protein array using the lithographic technique. (Lee et al., 2008. Reprinted with permission. (C) 2008 American Chemical Society)

8.7 Cell Patterning via Direct Protein Deposition

If the arrays are made in a millimeter scale, automated pipettors can be used, where a solution of protein or peptide can be dropwise added and evaporated on a surface (Fig. 8.13).

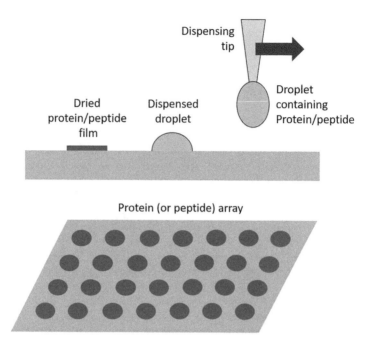

Fig. 8.13 Automated droplet dispenser system

8.8 Cell Patterning via Microcontact Printing

Glues (ECM proteins or RGD-containing peptides) can also be patterned using *microcontact printing* (*μCP*). In microcontact printing, a "stamp" is made using photolithography or soft lithography (refer to Chap. 6). Glue materials are added to this mold and "stamped" on the biomaterial surface. Mammalian cells are then added to create the desired cell pattern. Generally, a stamp is made from a photo-lithographed silicon wafer mold using polydimethylsiloxane (PDMS) (Fig. 8.14). This procedure is known as *PDMS replica molding*. This PDMS replica works as a "stamp," where the ink is a chemical ligand that accommodates glue protein patterning.

A self-assembled monolayer (SAM) of alkanethiol is used as an ink. We can treat the nonpatterned areas with passivating molecules that prevent the adhesion of glue protein. The glue protein is then added, for example, ECM protein such as fibronectin. Mammalian cells are then seeded and anchored only to the patterned area.

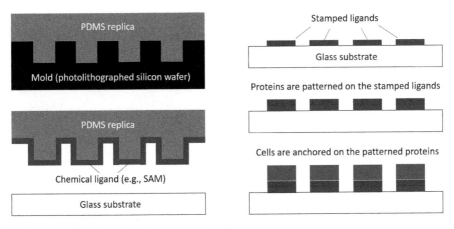

Fig. 8.14 Cell patterning via microcontact printing

8.9   Direct Cell Patterning via Inkjet Printing (Cell Printing)

While cell patterning has popularly been used toward printing cells in the desired and organized manner, it still requires "glue" such as ECM proteins and RGD-containing peptides to anchor the mammalian cells at the desired locations. There is an alternative way to "print" the cells directly without using such "glue."

Cell printing is a method to directly "print" cells in the desired pattern toward creating an organ or organ-like structure, that is, *organ printing*. Figure 8.15a shows an example of organ printing, where a ring of 10 printed spherical aggregates of cells (each ~500 μm diameter) within collagen gel. Each aggregate is printed with a special bioprinter. After 24 h, the spherical aggregates are connected to form a doughnut shape (Figure 8.15b). Repeating this unit structure can yield a honeycomb-shaped system (C at initial time point and D at 24 h). It is also possible to print two or more types of cells together, as shown in Fig. 8.16 (red = human fibroblasts; green = rat hepatocytes).

While we can use the automated droplet dispenser shown in the previous section (Fig. 8.13) to print cells directly, its resolution is quite limited, typically at millimeter scale, and inappropriate to print the patterns at μm scale.

In the past, *inkjet printers* have popularly been used to print the cells in the desired pattern directly. In inkjet printers, liquid inks in four different colors are sprayed through nozzles to print on paper. Inks are stored in *ink cartridges*. Some models use four separate ink cartridges, while some other model uses two ink cartridges – one for black and the other for three primary colors (cyan, magenta, and yellow). If we can replace the inks with different cell suspensions, it is possible to print up to four different cells on a substrate simultaneously. This parallel printing capability has caught the attention of many researchers around the world.

As the paper is not a preferred substrate for cells, plastic films (including overhead transparency films) have popularly been used. Unfortunately, many inkjet

Fig. 8.15 Example of
organ printing. (Rago
et al., 2009. Reprinted with
permission. (C) 2009 John
Wiley and Sons)

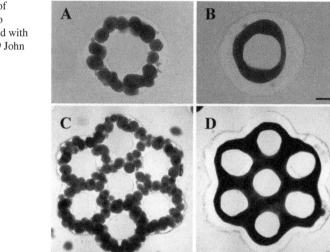

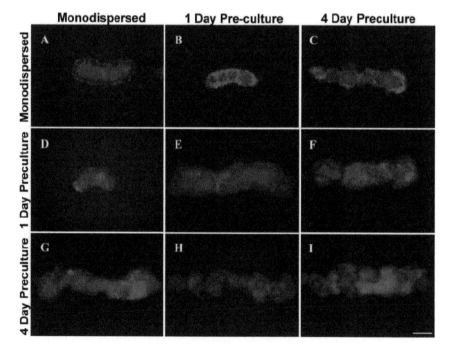

Fig. 8.16 Example of organ printing. (Rago et al., 2009. Reprinted with permission. (C) 2009
John Wiley and Sons)

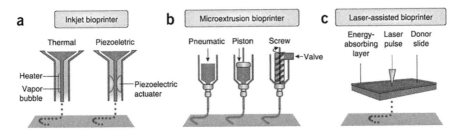

Fig. 8.17 Working principles of inkjet bioprinter and modern bioprinters for organ printing. (Murphy & Atala, 2014. Reprinted with permission. (C) 2014 Springer Nature)

printers use heat to dispense liquid inks through nozzles, which create complications to the mammalian cells. Therefore, efforts have been made to improve this principle further. A special bioprinter has been proposed, inspired by inkjet printing but not using the actual inkjet printer or cartridges. Figure 8.17 illustrates the concept of such a bioprinter compared to the inkjet printer, and Fig. 8.18 shows one commercial example. There exists a couple of cooling mechanisms that would prevent heat damage to the cells.

While we are discussing the 2D bioprinting, it is certainly possible to convert it into 3D bioprinting, through printing layer by layer. This topic will be further discussed in Chap. 9.

Question 8.2 Briefly and graphically explain (A) cell printing with nozzles (a.k.a. inkjet-based cell printing) and (B) cell patterning with printed protein spots. Compare the strengths and weaknesses of these two methods.

8.10 Laboratory Task 1: Contact Guidance on Microgroove

A simple method will be demonstrated in this task rather than a complicated process of creating micro- and nanogroove patterns on a synthetic biomaterial surface. A simple scalpel or a blade will be used to "cut" or "etch" a tissue culture plate surface.

Objective 1. Simple Fabrication of a Microgroove
1. Wear gloves and sterilize with ethanol.
2. Prepare a 24-well tissue culture plate.
3. Using a sterilized scalpel or a blade, cut or etch a microgroove into the well plate surface. If possible, try to vary the hand pressure applied to the scalpel or the blade to alter the channel depth and width. Also, try to create multiple lines in parallel or perpendicular (Fig. 8.19).

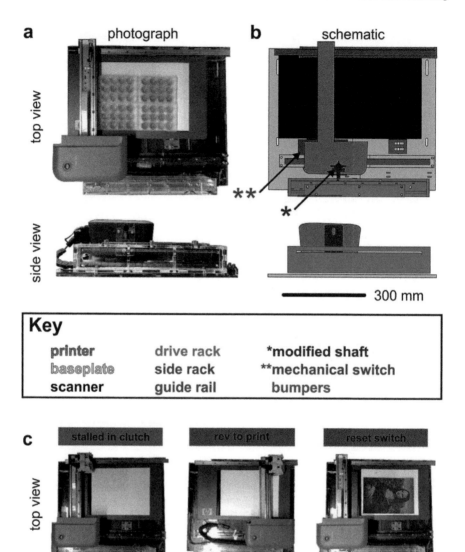

Fig. 8.18 An example of bioprinter inspired by inkjet printing technology. (Orloff et al., 2014. Reprinted with permission, (C) 2014 Royal Society of Chemistry)

Objective 2. Seeding Cells on a Microgroove

4. Prepare two wells, one without cut/etch (plain surface) and the other with cut/etch (microgroove surface). Sterilize the well plate with the lid open under UV light for 10 min.
5. Passage and resuspend a confluent cell culture in 4 mL fresh media (refer to Chap. 3).

Fig. 8.19 Etching a
24-well plate to create
microgroove patterns

6. Pipette 50 µL each of the cell suspension to the two wells. Allow droplets to stand undisturbed on the surface for 10 min. Do not shake or mix the well plate.
7. Fill each well with 250 µL of warmed, prepared media.

Objective 3. Fluorescence Imaging of Cells on a Microgroove

8. Stain the cells in each well following the protocol described in Chap. 7. DAPI staining is used to count the cell number, vinculin staining to confirm focal adhesion, and actin filament staining to verify cell morphology (rounded vs. spread).

Figure 8.20 shows the bright-field, DAPI-stained (nuclei), anti-vinculin and anti-IgG-FITC stained (vinculin), and phalloidin-TRITC stained (actin filaments) images of cells on a plain tissue culture plate surface. Cells exhibit spread morphology with plenty of vinculin presence, indicating a substantial amount of focal adhesion. Also, vinculins are found only where actin filaments are located.

Figure 8.21 shows the images of cells on a microgroove patterned surface. A bright-field image clearly shows an etched microgroove, which is quite rough. Such surface roughness will lower the surface energy and subsequently makes the surface hydrophobic, whereas the other area (tissue culture plate) should remain hydrophilic and cell friendly. As a result, cells can only be found just outside of the microgroove but not within the microgroove, as confirmed by the presence of nuclei, vinculins, and actin filaments. Also, cells near the microgroove are more aligned to the length of a microgroove.

Fig. 8.20 Bright-field, DAPI-stained (nuclei), anti-vinculin and anti-IgG-FITC stained (vinculin), and phalloidin-TRITC stained (actin filaments) images of cells on a plain tissue culture plate surface

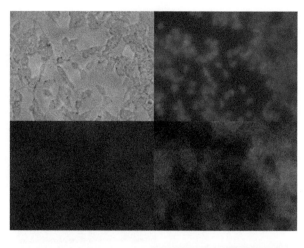

Fig. 8.21 Bright-field, DAPI-stained (nuclei), anti-vinculin and anti-IgG-FITC stained (vinculin), and phalloidin-TRITC stained (actin filaments) images of cells on a microgroove patterned surface

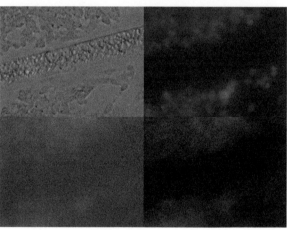

8.11   Laboratory Task 2: Cell Patterning via Droplet Collagen Deposition

A simple cell patterning method will be demonstrated in this task by depositing droplet(s) of a collagen solution, followed by cell seeding and proliferation on it.

9. Prepare a solution of 50 µg/mL rat tail collagen type I.
10. Dispense 10 µL droplet(s) within the 24-well plate. Let it dry under room temperature (preferably in a biosafety cabinet or a chemical hood) (Fig. 8.22).
11. Repeat Objectives 2 and 3.

Figure 8.23 shows the results, indicating that the cells can be found where the collagens are "printed" through droplet dispensing. The vinculin-stained images also show efficient focal adhesion.

Fig. 8.22 Dispensing the
droplets of a collagen
solution within a 24-well
plate

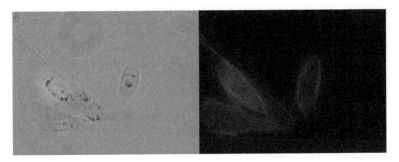

Fig. 8.23 Bright-field (left), and anti-vinculin and anti-IgG-FITC stained (right) images on a
collagen-printed surface via droplet dispensing

Review Questions

1. Define contact guidance. Why is it essential in tissue engineering?
2. What is the basement membrane? Why do we need to mimic basement membrane in tissue engineering?
3. Discuss the effects of the nanogroove width for cell adhesion and proliferation.
4. Focal adhesion can form between anchorage-dependent cells and the nanogrooves without the RGD peptide sequence. How can it happen?
5. How can micro- and nanogrooves provide flow-resistant adhesion?
6. Compare cell patterning methods with versus without using proteins or peptides.
7. Discuss the strengths and weaknesses of the automated droplet dispenser system toward cell patterning.
8. Describe cell patterning by lithographic printing.
9. Describe cell patterning by microcontact printing.

10. What are cell printing and organ printing? How are they different from conventional cell patterning methods?
11. What are the benefits and issues of inkjet bioprinting using commercial inkjet printers? Why is a dedicated inkjet bioprinter necessary?

References

Abrams, G. A., Schaus, S. S., Goodman, S. L., Nealey, P. F., & Murphy, C. J. (2000). *Cornea, 19*(1), 57–64. https://journals.lww.com/corneajrnl/Fulltext/2000/01000/Nanoscale_Topography_of_the_Corneal_Epithelial.12.aspx

Gamboa, J. R., Mohandes, S., Tran, P. L., Slepian, M. J., & Yoon, J. Y. (2013). Linear fibroblast alignment on sinusoidal wave micropatterns. *Colloids and Surfaces B: Biointerfaces, 104*, 318–325. https://doi.org/10.1016/j.colsurfb.2012.11.035

Lee, J. Y., Shah, S. S., Zimmer, C. C., Liu, G. Y., & Revzin, A. (2008). Use of photolithography to encode cell adhesive domains into protein microarrays. *Langmuir, 24*, 2232–2239. https://doi.org/10.1021/la702883d

Loesberg, W. A., Walboomers, X. F., van Loon, J. J. W. A., & Jansen, J. A. (2007). Simulated microgravity activates MAPK pathways in fibroblasts cultured on microgrooved surface topography. *Cell Motility and the Cytoskeleton, 65*, 116–129. https://doi.org/10.1002/cm.20248

McCracken, K. E., Tran, P. L., You, D. J., Slepian, M. J., & Yoon, J. Y. (2013). Shear- vs. nanotopography-guided control of growth of endothelial cells on RGD-nanoparticle-nanowell arrays. *Journal of Biological Engineering, 7*, 11. https://doi.org/10.1186/1754-1611-7-11

Murphy, S. V., & Atala, A. (2014). 3D bioprinting of tissues and organs. *Nature Biotechnology, 32*, 773–785. https://doi.org/10.1038/nbt.2958

Orloff, N. D., Truong, C., Cira, N., Koo, S., Hamilton, A., Choi, S., Wu, V., & Riedel-Kruse, I. H. (2014). *RSC Advances, 4*, 34721–34728. https://doi.org/10.1039/C4RA05932H

Rago, A. P., Dean, D. M., & Morgan, J. R. (2009). Controlling cell position in complex heterotypic 3D microtissues by tissue fusion. *Biotechnology and Bioengineering, 102*, 1231–1241. https://doi.org/10.1002/bit.22162

Tran, P. L., Gamboa, J. R., McCracken, K. E., Riley, M. R., Slepian, M. J., & Yoon, J. Y. (2013). Nanowell-trapped charged ligand-bearing nanoparticle surfaces – A novel method of enhancing flow-resistant cell adhesion. *Advanced Healthcare Materials, 2*, 1019–1027. https://doi.org/10.1002/adhm.201200250

Chapter 9
3D Scaffold Fabrication

In the previous chapters, we have learned how biomaterial surfaces, mainly in 2D and thin sheets mimicking basement membranes in the human body, can be modified to accommodate focal adhesion and eventually migration, differentiation, and proliferation toward successful tissue development. In practical applications, however, other types of tissues are not in a 2D sheet shape.

Hydrogels have already been discussed in the previous chapter as 3D scaffold materials for tissue engineering. However, their porous structures are pretty randomly distributed, and there is hardly any "engineering" control over their designs.

Two methods have popularly been adopted and investigated toward fabricating 3D tissue-engineered scaffolds: electrospinning and 3D printing. This chapter will learn the basic principles of these two methods and utilize them to fabricate 3D tissue-engineered scaffolds.

Inquiry 1. Have you heard about electrospinning or electrospun micro- or nanofibers?

Inquiry 2. Have you heard about 3D printing? Can you describe how it works? How can it be used toward fabricating a 3D tissue-engineered scaffold?

9.1 3D Scaffolds

In previous chapters, we have learned how to modify the biomaterial surfaces to accommodate mammalian cells' focal adhesion toward their successful migration, differentiation, and proliferation. We focused our discussions on 2D, sheet-like biomaterials that can mimic basement membranes typically found in natural tissues. While this approach has worked to construct various types and shapes of tissue constructs, it is inherently limited as it does not offer a proper 3D structure.

© Springer Nature Switzerland AG 2022
J.-Y. Yoon, *Tissue Engineering*, https://doi.org/10.1007/978-3-030-83696-2_9

Fig. 9.1 A 3D collagen
gel serves as a 3D
tissue-engineered scaffold

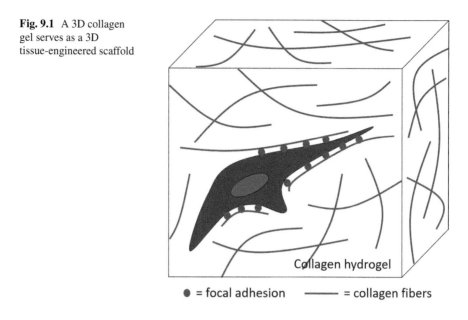

● = focal adhesion   ——— = collagen fibers

A better approach is to create a 3D scaffold made from either synthetic or natural materials. *Hydrogels* have already been suggested and have shown numerous successes in the past. However, their structures cannot be "engineered" as their porous structures are formed quite in a random manner (Fig. 9.1).

9.2   Electrospinning

Electrospinning is a method to create a network of polymer fibers from the polymers dissolved in a solvent. Figures 9.2 and 9.3 show the schematic illustration of the electrospinning process. The polymer solution is loaded into a syringe. A *syringe pump* is a device that can push the plunger of a syringe at quite a slow rate, which is the crucial component in the electrospinning process. Pushing the plunger of a syringe at a slow rate will create a stream of the polymer solution, and upon evaporation of the solvent, it will leave the polymer fibers. As the inner diameter of a syringe needle is typically in the sub-millimeter scale range, the resulting fibers should also have a few hundreds of micrometer (= sub-millimeter) scale. Through applying an electrical voltage to these fibers, you can elongate the polymer fibers to have a few or a few tens of micrometer, that is, typical sizes of collagen fibers in the extracellular matrix (ECM). By further increasing the voltage, we can also produce sub-micrometer fibers, that is, nanofibers. Voltage is applied to a syringe needle (typically made from various metals), and a metal plate is placed away from the syringe, called a *collector*. While a planar sheet is generally used as a collector in the electrospinning process (Fig. 9.2), it is also possible to have a cylindrical

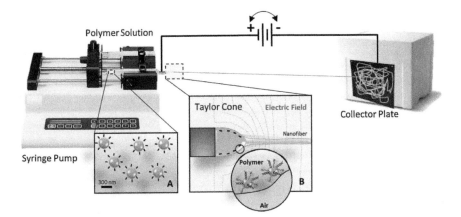

Fig. 9.2 Electrospinning on a plate. (Nicolini et al., 2016. Reprinted with permission. (C) 2016 Elsevier)

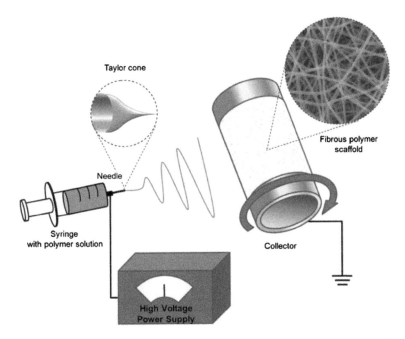

Fig. 9.3 Electrospinning on a rotating cylinder. (Rim et al., 2013. Reprinted with permission. (C) 2013 IOP Publishing Ltd.)

collector shown in Fig. 9.3 to create a cylindrical scaffold. Such cylindrical collector is slowly rotated to provide uniform distribution of fibers. The resulting fibers are a randomly oriented network of fibers, as shown in the inset of Fig. 9.3, and they look like spaghetti noodles. Depending on the size of fibers, they are called *electrospun microfibers* or *electrospun nanofibers*.

Electrospinning has been very popular for various applications. However, its tissue engineering applications make up only a fraction of them. It has always been popular as the method is relatively easy. Most importantly, it requires only two rather inexpensive laboratory equipment typically available in many research laboratories – a syringe pump and a high-voltage power supply.

9.3   Materials Selection for Electrospinning Toward Tissue Engineering

Electrospinning has popularly been conducted using *polycaprolactone (PCL)*. Its chemical structure is shown in Fig. 9.4. PCL belongs to the category of *polyester*, and because of that, it has a highly elastic material property, which is desirable for constructing soft tissue mimics. It is also biodegradable, which is a good advantage for tissue engineering applications. However, PCR biodegrades relatively slow, often take more than 2 years to be fully biodegraded.

Better alternatives are *poly(lactic acid) (PLA)*, *poly(glycolic acid) (PGA)*, and their copolymer *PLGA*, as shown in Fig. 9.5. They have already been explained in Sect. 6.9 and Fig. 6.9. They have superior biocompatibility and biodegradability over PCL. They also belong to the polyester family, as shown in Fig. 9.5, and thus have highly elastic material properties. Specifically, PLGA biodegrades faster than all others, and we can even adjust its mechanical property by controlling the copolymer ratio.

All these materials (PCL, PLA, PGA, PLGA, etc.) have successfully been used for constructing electrospun micro- and nanofibers. As expected, a copolymer of PCL and PLA has also been tested.

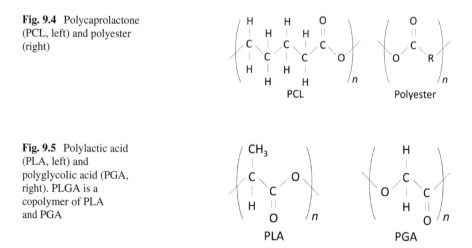

Fig. 9.4 Polycaprolactone (PCL, left) and polyester (right)

Fig. 9.5 Polylactic acid (PLA, left) and polyglycolic acid (PGA, right). PLGA is a copolymer of PLA and PGA

Natural polymers have also been used to create electrospun micro- and nanofibers, most notably collagen fibers and *gelatin* (denatured collagen fibers). *Silk* (protein fibers from silkworms or spiders) is another popular material for electrospun fibers. Likewise, *chitin* (polysaccharides from crabs and shrimps) has been successfully electrospun into micro- and nanofibers.

9.4 Modification of Electrospun Fibers for Tissue Engineering

Electrospun fibers are generally biocompatible and biodegradable. However, they cannot form focal adhesion with cells due to the lack of RGD peptide sequence. Therefore, the fiber surface must be modified to accommodate focal adhesion. For example, nanostructures can be imprinted on electrospun fibers' surface. Such imprinting can accommodate focal adhesion without the need for RGD sequence. (Focal adhesion on nanostructures was explained in Sect. 8.3 and Figs. 8.6 and 8.7.) Chemical ligands (mimicking RGD) can be added to the polymer to accommodate focal adhesion. ECM proteins, such as collagen, fibronectin, vitronectin, and laminin, can be physically adsorbed on the fiber surface to accommodate focal adhesion. Covalent conjugation of such ECM proteins is also possible, typically modifying the polymer fiber with carboxyl groups and creating covalent peptide bonds between these carboxyl groups on the polymer fiber and the amine groups of the ECM proteins. *Carbodiimides*, usually in the form of *1-ethyl-3-(3-dimethylaminopropyl)carbodiimide* (*EDC* or *EDAC*), are typically used to achieve such covalent conjugation. These methods are summarized in Fig. 9.6.

9.5 Contact Guidance on Electrospun Fibers

While most electrospun fibers are randomly oriented, it is also possible to make the fibers to be aligned in parallel. With proper surface modification, cells can be aligned to the direction of fibers, for example, contact guidance, to fit various tissues' requirements, as shown in Fig. 9.7. For example, electrospun fibers can be aligned in one direction to resist the bulk liquid flow (e.g., blood vessels). They can also be aligned to resist tensile or compressional force applied in one direction (e.g., tendons). Alternatively, they can be aligned in random directions to ensure elasticity in multiple directions (e.g., cardiac tissues). The diameter of electrospun fibers (micro- vs. nano-) is also essential in achieving proper adhesion strength, as previously explained in Sect. 6.2 and Fig. 6.3.

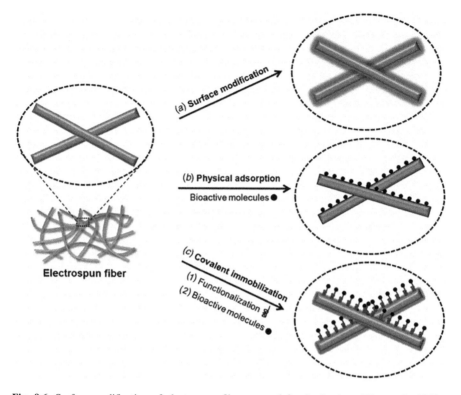

Fig. 9.6 Surface modification of electrospun fibers toward focal adhesion. (Rim et al., 2013. Reprinted with permission. (C) 2013 IOP Publishing Ltd.)

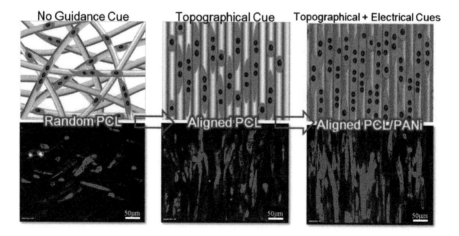

Fig. 9.7 Cells are aligned via contact guidance on randomly oriented electrospun fibers (left), parallelly aligned electrospun fibers (middle), the same with electrical cues (right). (Chen et al., 2013. Reprinted with permission. (C) 2013 Elsevier)

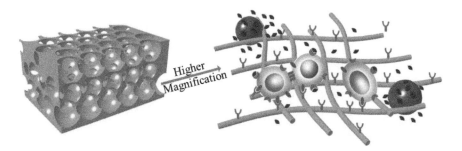

Fig. 9.8 Controlled release of growth factors and bioactive factors from electrospun fiber scaffolds. (Zhang et al., 2012. Reprinted with permission. (C) 2012 Elsevier)

9.6   Controlled Release of Growth Factors and Bioactive Factors from Electrospun Fibers

To make the cells on electrospun fibers continuously proliferate and perform appropriate functions, we should make the electrospun fiber scaffold release growth factors and other bioactive factors in a controlled manner. Slow and controlled release of such factors and drugs from a matrix is called the *controlled release of drugs*, or *drug delivery*, another important aspect of tissue engineering. Figure 9.8 shows the example of such controlled releases. Cells are anchored on electrospun fibers through cell-binding ligands immobilized on fibers (ECM proteins, RGD peptide, etc.) and cell adhesion receptors (integrins). Particles are also "trapped" within the electrospun fibers network, which slowly releases bioactive factors or growth factors (thus acting as drug delivery vehicles). They can be porous microparticles where such factors are incorporated within their pores (the molecular diffusion controls the release through the pores). Alternatively, growth factors or bioactive factors can also be incorporated into the fibers and slowly released.

9.7   Core–Shell Structured Electrospun Fibers

Electrospun fibers can also be made from two different polymers, one at the core and the other at the shell, to take advantage of both. Usually, the polymer with the better mechanical property, but the insufficient surface property is placed in the core – the one with the better surface property but the insufficient mechanical property at the shell. For example, PCL can be used at the core side, providing a better mechanical property (than PLGA), and PLGA at the shell side, providing a better biocompatibility (than PCL). Also, the core can be made from a solution of biomolecules (growth factors, bioactive factors, etc.) that can be slowly eluted through the shell-side polymers, as shown in Fig. 9.9. It is also possible to create multilayered electrospun fibers that can mimic natural tissue's complex structure.

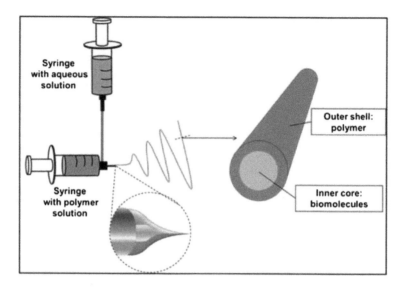

Fig. 9.9 Core–shell structured electrospun fiber with growth factor or bioactive factor incorporated with its pore. (Rim et al., 2013. (C) 2013 IOP Publishing Ltd.)

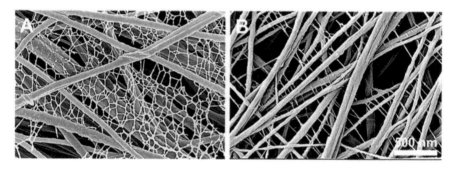

Fig. 9.10 Nanonets are added to electrospun fibers to mimic native ECM better. (Saudi et al., 2020. (C) 2020 Royal Society of Chemistry)

9.8   Addition of GAG-Like Structure to Electrospun Fibers

As explained in Sect. 6.1 and Fig. 6.1, collagen fibers make the major backbone of ECM, while glycosaminoglycans (GAGs) interweave the collagen fibers. While the diameters of collagen fibers are a few to a few tens of micrometer, those of GAGs are in the nanometer scale. The addition of a GAG-like structure to electrospun fibers would undoubtedly bring the scaffold close to the native ECM. Figure 9.10 shows one such attempt, where nanonets are added to the electrospun fibers.

9.9   3D Printed Scaffolds

3D printing may be the ultimate solution in creating a tissue engineering scaffold in whatever shape we want. The scaffold can be designed using *computer-aided design* (*CAD*) software, such as *AutoCAD* (from Autodesk) or *SolidWorks* (from Dassault Systèmes). We can take the scaffold designs from existing drawings for a human organ or 3D imaging (e.g., MRI or CT) taken from a specific patient (patient-specific). The CAD design is sliced into a stack of 2D images, called an *STL file* (representing *stereolithography*). And a 3D printer will print the 3D scaffold, layer by layer, using the STL file. Figure 9.11 shows the 3D image of a human skull with a damaged area (top left). A CAD design is developed to "patch" the damaged area (top right). The 3D printer then "prints" this design layer by layer using the STL file.

Like electrospinning, 3D printed scaffolds generally receive a surface modification. This modification includes coating with ECM proteins or RGD peptides to promote focal adhesion. Porosity can also be added to ensure mechanical interlocking with the cells, focal adhesion (without the need for ECM proteins or RGD peptides), and loading with growth or bioactive factors. While the 3D printed material can directly be added to the human body as an implant, it is desirable to grow cells in vitro and implant later, that is, as a tissue-engineered transplant.

While there are numerous methods of 3D printing, we will discuss only the methods relevant to tissue engineering in this chapter. Specifically, 3D printing methods that can generate porous structures are beneficial for constructing tissue engineering scaffolds. They can generate inherent micro- and nanostructures favorable for focal adhesion and tissue ingrowth, and growth/bioactive factors can be incorporated within the scaffolds.

Fig. 9.11 3D printed scaffold. (Lee et al., 2005. Reprinted with permission. (C) 2005 Elsevier)

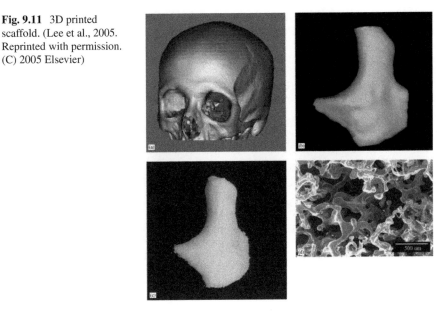

9.10   Various 3D Printing Methods

In *powder bed and inkjet head 3D printing*, also known as *binder jet 3D printing*, an inkjet printer prints the object layer by layer, where the powder is continuously fed into a bed (Fig. 9.12). Once 3D printing is complete, powders are removed, leaving only the object. When it was first conceived in the 1990s, they called this method simply 3D printing (3DP) and trademarked it. As many other 3D printing methods have emerged later, 3D printing now represents all methods collectively, where the original method is now distinguished as powder bed and inkjet head 3D printing.

A variety of materials can be used for powder bed and inkjet head 3D printing. Polymers used for electrospinning – PCL, PLA, and PLGA – have successfully been used. Also, polysaccharides such as starch, dextran, hyaluronic acid (one of the GAGs), and alginate have been used. Calcium phosphates have also been used, including tricalcium phosphate [TCP; $Ca_3(PO_4)_2$] and hydroxyapatite [HA; $Ca_5(PO_4)_3(OH)$], which are beneficial for bone tissue engineering. Even ECM proteins can be 3D printed using this method, including fibronectin, collagen, and gelatin. It is also possible to directly print mammalian cells.

In the *fused deposition modeling (FDM)*, *thermoplastic materials* (polymers become flexible and thus moldable in high temperature) are extruded through a heated nozzle to conduct layer-by-layer printing, as shown in Fig. 9.13. Materials typically come in filaments (*3D printing filaments*) in rolls, with a few millimeter diameter. Similar to color laser printers or color inkjet printers, multiple filaments can be used simultaneously. Two filaments are commonly used, one as a building material and the other as a support material. The support material is later dissolved in water to create a complex feature. A low-end FDM 3D printer can print only one material (build material) and is thus incapable of printing complex features. FDM is probably the most well-known and widely used 3D printing method available at the time of writing. Patent on FDM expired some time ago, and the cost of FDM 3D

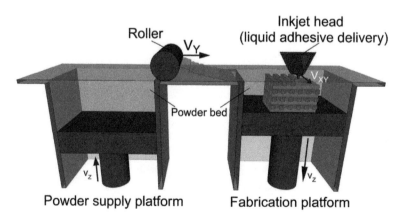

Fig. 9.12 Powder bed and inkjet head 3D printing. (Billiet et al., 2012. Reprinted with permission. (C) 2012 Elsevier)

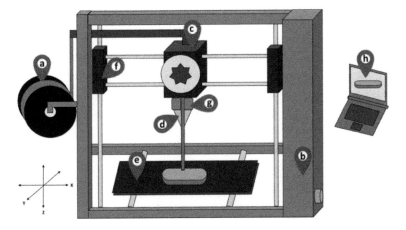

Fig. 9.13 Fused deposition modeling (FDM) 3D printer: (**a**) a roll of printing materials, (**b**) enclosure, (**c**) extruder head, (**d**) nozzle, (**e**) platform, (**f**) motor, (**g**) heater, and (**h**) CAD software. (Araujo et al., 2019. (C) 2019 Araujo et al. Open access article distributed under the terms of the Creative Commons Attribution 4.0 International License)

printer has dramatically been reduced from previously a few tens of thousands of US dollars to only a few hundreds of US dollars.

Objects created by FDM 3D printers are also highly porous, which may be a good character toward tissue engineering scaffold. They also exhibit good mechanical strength. However, to "mold" the polymers into the desired shape, high temperature must be used, and the incorporation of biological materials directly into the FDM 3D printing can be difficult. Besides, the choice of materials is also limited as they must be thermoplastic.

The following is a list of materials that can be used for FDM 3D printing. Thermoplastic materials have two critical temperatures that can define their material property and moldability: (1) *glass transition temperature* or T_g, where the material changes from "glassy" to "rubbery" state where it becomes pliable, and (2) *melting temperature* or T_m, where the material is melt into liquid form. While it is possible to bend the material on or above T_g, materials must be heated beyond T_m to be extruded through the nozzle in liquid form. Most materials have T_m from 200 °C to 300 °C, which is the desired operating condition of FDM 3D printing. As expected, FDM 3D printing cannot directly print biological materials such as ECM proteins as they would completely lose their shape and function when melted. While ABS has been the most popular choice due to its excellent material property, PLA and PLGA have been the most popular choice for tissue engineering scaffolds due to their biocompatibility and biodegradability (Table 9.1).

In *stereolithography* (*SLA*) 3D printing, HeCd laser cures (i.e., cross-links) the polymer liquid in 2D patterns. The area not exposed to the laser will not be cured (cross-linked) and drained. The procedure is repeated layer by layer to create a 3D object (Fig. 9.14). Unlike the other methods, materials are not "printed." Instead, they are slowly emerging from the liquid bath upon laser irradiation. As a result,

Table 9.1 Materials used for FDM 3D printing, shown with a glass transition temperature (T_g) and melting temperature (T_m)

	T_g (°C)	T_m (°C)
ABS, acrylonitrile-butadiene-styrene copolymer	105	220–230
PVC, poly(vinyl alcohol)	83	212–310
PET, poly(ethylene terephthalate)	69–77	265–284
PLGA, poly(lactic-co-glycolic acid)	40–60	223–260
Nylon	30–75	177–272
PCL, polycaprolactone	−60	60

Fig. 9.14 Stereolithography (SLA) 3D printing. (Huang et al., 2020. (C) Huang et al. Open access article distributed under the terms of the Creative Commons Attribution 4.0 International License)

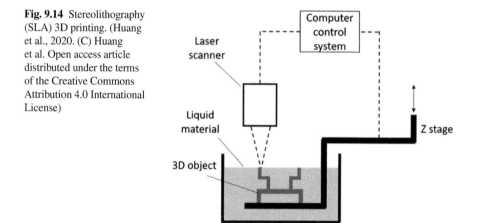

complex shapes with internal architecture can easily be printed using this method (without the need for support material), and very high resolution (down to 1 μm) can be achieved.

As this method is based on the material's ability to be cured (cross-linked), highly reactive polymers that have double bonds in the backbone, such as acrylics and poly(propylene fumarate) (PPF), as well as epoxies, have been used in SLA 3D printing. Recently, regular polymers such as PEG [poly(ethylene glycol)], PCL, and PLA have also been used, although modifications are necessary to these polymers to induce laser-induced curing (cross-linking). Therefore, the choice of materials is greatly limited with this method. Also, the resulting cross-linked polymers do not always show the best material property.

Selective laser sintering (SLS) 3D printing is a method similar to powder bed and inkjet head 3D printing. A CO_2 laser is used to coalesces the powder (= sintering) into solid to create a 2D pattern. The laser is heated only above T_g but not over T_m. [If the temperature is heated beyond T_m, it becomes *selective laser melting (SLM.)*]

This procedure is repeated layer by layer to create a 3D construct. Unbound and loose powder is removed. Like powder bed and inkjet head 3D printing, various materials can be used, including PCL, polyether ether ketone, tricalcium phosphate, hydroxyapatite, titanium, etc. However, its material choice is still limited compared to the powder bed and inkjet head 3D printing as the material must be sintered by CO_2 laser, although it is still better than SLA and FDM. It also takes the advantages of SLA – no need for support material and high resolution. However, the powder's size limits the resolution, which is around 10 μm (still better than FDM). The real problem of SLS is the unwanted incorporation of powders into the final construct, and such "trapped" powder must be removed.

9.11 Hydrogel Bioprinting

In most 3D printing methods, nozzles at high temperatures or lasers are used to print the synthetic or natural polymers layer by layer. In *hydrogel bioprinting*, on the other hand, natural biomaterials in water (i.e., hydrogels) are extruded through a nozzle to print the hydrogels layer by layer to construct a 3D object. It can be considered a subset of inkjet bioprinting described in Sect. 8.9. The hydrogel materials used for bioprinting include (1) natural hydrogels such as *agar* (agarose-based hydrogel, where *agarose* is a polysaccharide from seaweed), (2) ECM protein-based hydrogels such as *gelatin* (collagen-based hydrogel), (3) chitin (polysaccharides from crabs and shrimps), (4) tricalcium phosphate (TCP; toward bone tissue engineering), and (5) PLGA (biodegradable polymer). Other agarose- and collagen-based hydrogels can also be used. Figure 9.15 shows the schematic illustration of hydrogel bioprinting. Many repeating units from hydrogel bioprinting can be assembled to create a large structure.

The primary difference of hydrogel bioprinting from other 3D printing methods is the use of hydrogels, where the network of natural polymers creates a spongy-like structure, mostly filled with water (over 80–90%). This structure closely mimics the

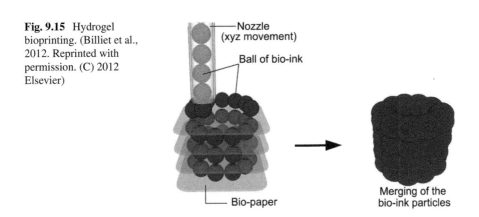

Fig. 9.15 Hydrogel bioprinting. (Billiet et al., 2012. Reprinted with permission. (C) 2012 Elsevier)

natural ECM, and cells can be incorporated "within" the scaffold (made from hydrogels), which is not possible with other 3D printing methods. Also, printing can be conducted at room temperature, which is a benefit in printing natural biomaterials. Finally, any polymers that can be made into hydrogels can be bioprinted.

However, such a hydrogel nature is also hydrogel bioprinting's significant weaknesses. As they are hydrogels with spongy-like structures, the material is not stiff and may collapse easily. To preserve its shape, the densities of polymer and liquid medium (water) must be similar. Thus, it is challenging to fabricate complex shapes, and the printing resolution is low.

9.12   Adding Porosity and Nanostructure to 3D Printing

Toward tissue-engineered scaffolds, it is sometimes necessary to add porosity to the final construct. While many 3D printed constructs are inherently porous, it is required to create porous structures whose sizes are comparable to those of mammalian cells – e.g., a few tens of micrometer. Popular 3D printing methods such as FDM have 100–200 μm resolutions, which is insufficient. Oxygen transport is limited to 200 μm in most tissues (this will be covered in the next chapter), and 3D printed scaffolds must be printed at a resolution far better than this limit. Such additions are typically done using *porogens*, which generally are NaCl powder or PLGA particles. Once porogens are added to the materials and the object is 3D printed, porogens can later be leached out, creating porous structures (Fig. 9.16). The size of

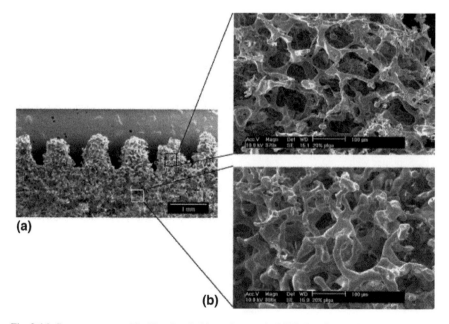

Fig. 9.16 Porous structured in 3D printed object. (Lee et al., 2005. Reprinted with permission. (C) 2005 Elsevier)

added porogens dictates the size of such pores. However, porogens must be smaller than the thickness of layers. Porogens can easily be added to powder bed and inkjet head 3D printing and SLS 3D printing, and mixed with the powders that those methods use.

We can sometimes add porosity to the surface of 3D printed scaffolds by introducing nanoarchitectures. Like electrospinning, it may also be necessary to pattern growth factors and other bioactive factors onto the 3D printed scaffold.

9.13   Indirect 3D Printing

It may be necessary to use an in vivo-like material for 3D printing. In such a case, a mold can be 3D printed, and a desired (biodegradable) polymer can be added into the mold, cured, and removed from the mold to obtain the desired scaffold. This procedure is called *indirect 3D printing.*

The most apparent advantage of indirect 3D printing is its more extensive choice of materials. Besides, the size of porogens is no longer limited by the layer thickness. However, a complex feature cannot be duplicated from the mold as the fabrication is not dictated by laser and the support material is not available.

9.14   Laboratory Task 1: Electrospinning

In this task, electrospun fibers are constructed using PCL, and their material properties are confirmed with microscopy and contact angle analysis.

Objective 1. Preparation of Electrospinning

1. Fold a 10 cm × 10 cm aluminum foil. Take 1 cm × 1 cm microscope coverslips and place multiple coverslips on the aluminum foil using double-sided tapes.
2. Place the foil, with the coverslips facing toward the syringe pump (Fig. 9.17), on the copper plate. Set the distance between the needle (of a syringe pump) and the foil at 9 cm. Increasing the distance will result in thinner electrospun fibers.
3. Dissolve 0.1 g PCL in 1 mL of 1,1,1,3-hexafluoro-2-propanol (HFIP). Vortex for 3 min. Load it into a 10 mL syringe and 23-gauge needle tip. Push the plunger to remove any air trapped in a syringe.
4. Load the syringe onto the syringe pump and lock it into place.
5. Prepare a DC power supply capable of generating 15 kV.
6. Clamp the ground wire of a DC power supply onto the copper plate.
7. Clamp the high-voltage wire of a DC power supply onto the end of the syringe needle.

Fig. 9.17 Electrospinning apparatus

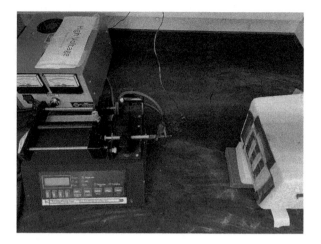

Objective 2. Electrospinning

8. Set the syringe pump rate to 20 µL/min. You will need to know the syringe diameter and height to determine the pump rate.
9. Turn the DC power source to 15 kV (15,000 V). Avoid contact with a power source, wiring, and any other conductive materials to prevent injuries. To make any changes to the system, you must turn off the power source.
10. Turn the syringe pump on by pressing the Pump button.
11. To create electrospun fibers in all coverslips, you may need to readjust a dock's positions (with the DC power voltage off).
12. Continue the process until you notice a formation of uniform, dense layers of fibers in each coverslip.
13. Turn off the power supply, then the syringe pump. Cool down for 2–3 min.
14. Unclamp the positive wire from the needle tip.
15. Using a scalpel or a blade, carefully cut the fibers in between the coverslips. Be careful not to disturb the fibers on each coverslip.
16. Remove coverslips. Place each coverslip in a petri dish and cover the lid.

Objective 3. Characterization of Electrospun Fibers

17. Using laboratory task 2 of Chap. 6 (Sect. 6.12), measure the electrospun fiber substrate's contact angle.
18. Using laboratory task 3 of Chap. 6 (Sect. 6.13), measure the root mean square roughness of electrospun fiber substrate.

The image of a water droplet on a PCL electrospun fiber is shown in Fig. 9.18. The contact angle is evaluated as 74°, which is slightly lower than that of pure PCL. While PCL itself is hydrophobic (contact angle = 83°), the presence of HFIP (alcohol; polar solvent and thus hydrophilic) slightly lowers the contact angle.

Figure 9.19 shows the bright-field microscopic image of a PCL electrospun fiber substrate, indicating a spaghetti-like fiber structure.

Fig. 9.18 Contact angle on a PCL electrospun fiber substrate

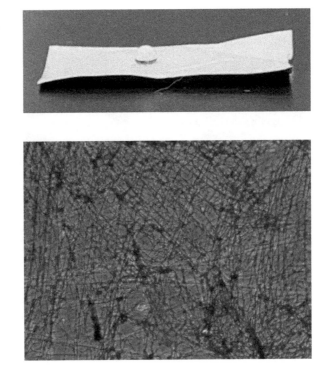

Fig. 9.19 Bright-field microscopic image of a PCL electrospun fiber substrate

9.15 Laboratory Task 2: 3D Printing

In this task, a 3D scaffold will be printed using a low-cost FDM 3D printer, following the method described in Mohanty et al., *Materials Science and Engineering C*, Volume 55, Pages 569–578, 2015, https://doi.org/10.1016/j.msec.2015.06.002. This article is available in open access.

19. Using appropriate CAD software, design the woodpile stacked shape of a scaffold (Fig. 9.20). While the authors have constructed the scaffold (mold) using PVA, it is okay to use a different material, for example, PLA or PLGA.
20. Set the 3D printing parameters to:

    Layer height = 0.2 mm (200 μm; typical for FDM 3D printing).
    Infill pattern = woodpile (or hexagonal).
    Nozzle temperature = 200 °C for PVA (change as necessary for PLA or PLGA).
    Build platform temperature = 40 °C.
    Feed rate = 20 mm/s.

21. Identify contact angle and capture a bright-field light microscopic image of the scaffold (Fig. 9.21).

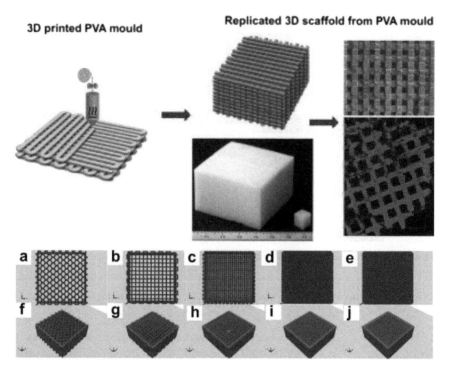

3D printed PVA mould

Replicated 3D scaffold from PVA mould

Fig. 9.20 3D printed scaffold. (Mohanty et al., 2015. (C) 2015 Mohanty et al. Open access article distributed under the terms of the Creative Commons Attribution 4.0 International License)

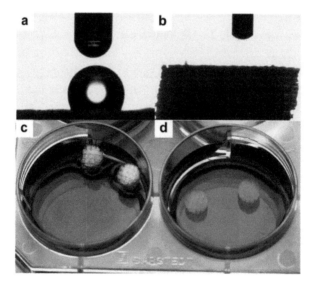

Fig. 9.21 Top: water contact angle measurement on a 3D printed scaffold. Bottom: the resulting 3D printed scaffold is placed in a hepatocyte cell culture, where hepatocytes are anchored to a 3D printed scaffold's surface and proliferate on it (Mohanty S, Larsen LB, Trifol J, Szabo P, Burri HVR, Canali C, Dufva M, Emneus J, Wolff A. 2015. Fabrication of scalable and structured tissue engineering scaffolds using water dissolvable sacrificial 3D printed moulds. *Materials Science and Engineering C* 55: 569–578. https://doi.org/10.1016/j.msec.2015.06.002. (C) 2015 Mohanty et al. Open access article distributed under the terms of the Creative Commons Attribution 4.0 International License)

Review Questions

1. What is hydrogel?
2. Describe how electrospinning works. Why is electrospinning popular toward tissue engineering (TE) scaffolds?
3. Compare the strengths and weaknesses of PCL, PLA/PGA/PLGA, and natural polymers for electrospinning toward TE scaffolds.
4. Why should the surfaces of electrospun fibers be modified toward TE scaffolds? How can such modifications be made?
5. Compare random versus aligned orientations of electrospun fibers toward TE applications.
6. Why should the growth factors or bioactive factors be incorporated into the electrospun fibers toward TE scaffolds? How can such additions be made?
7. Why do you need core–shell structured electrospun fibers for TE applications?
8. What are the benefits of adding GAG-like structure to the electrospun fibers for TE applications?
9. Compare powder bed and inkjet head, FDM, SLA, and SLS 3D printing, as well as bioprinting. Discuss their strengths and weaknesses.
10. List the materials that can be used for powder bed and inkjet head, FDM, SLA, and SLS 3D printing, as well as bioprinting.
11. What makes hydrogel bioprinting distinctly different from other 3D printing methods?
12. Why do you need to add porosity or nanostructure to 3D printed scaffold?
13. What is porogen for 3D printing?
14. What is indirect 3D printing?

References

Araujo, M. R. P., Sa-Barreto, L. L., Gratieri, T., Gelfuso, G. M., & Cunha-Filho, M. (2019). The digital pharmacies era: How 3D printing technology using fused deposition modeling can become a reality. *Pharmaceutics, 11*, 128. https://doi.org/10.3390/pharmaceutics11030128

Billiet, T., Vandenhaute, M., Schelfhout, J., van Vlierberghe, S., & Dubruel, P. (2012). A review of trends and limitations in hydrogel-rapid prototyping for tissue engineering. *Biomaterials, 33*, 6020–6041. https://doi.org/10.1016/j.biomaterials.2012.04.050

Chen, M. C., Sun, Y. C., & Chen, Y. H. (2013). Electrically conductive nanofibers with highly oriented structures and their potential application in skeletal muscle tissue engineering. *Acta Biomaterialia, 9*, 5562–5572. https://doi.org/10.1016/j.actbio.2012.10.024

Huang, J., Qin, Q., & Wang, J. (2020). A review of stereolithography: Processes and systems. *PRO, 8*, 1138. https://doi.org/10.3390/pr8091138

Lee, M., Dunn, J. C. Y., & Wu, B. M. (2005). Scaffold fabrication by indirect three-dimensional printing. *Biomaterials, 26*, 4281–4289. https://doi.org/10.1016/j.biomaterials.2004.10.040

Mohanty, S., Larsen, L. B., Trifol, J., Szabo, P., Burri, H. V. R., Canali, C., Dufva, M., Emneus, J., & Wolff, A. (2015). Fabrication of scalable and structured tissue engineering scaffolds using water dissolvable sacrificial 3D printed moulds. *Materials Science and Engineering C, 55*, 569–578. https://doi.org/10.1016/j.msec.2015.06.002

Nicolini, A. M., Toth, T. D., & Yoon, J. Y. (2016). Tuneable nanoparticle-nanofiber composite substrate for improved cellular adhesion. *Colloids and Surfaces B: Biointerfaces, 145*, 830–838.
https://doi.org/10.1016/j.colsurfb.2016.05.079

Rim, N. G., Shin, C. S., & Shin, H. (2013). Current approaches to electrospun nanofibers for tissue
engineering. *Biomedical Materials, 8*, 014102. https://doi.org/10.1088/1748-6041/8/1/014102

Saudi, S., Bhattarai, S. R., Adhikari, U., Khanal, S., Sankar, J., Aravamudhan, S., & Bhattarai,
N. (2020). Nanonet-nano fiber electrospun mesh of PCL-chitosan for controlled and extended
release of diclofenac sodium. *Nanoscale, 12*, 23556–23569. https://doi.org/10.1039/
D0NR05968D

Zhang, Z., Hu, J., & Ma, P. X. (2012). Nanofiber-based delivery of bioactive agents and stem
cells to bone sites. *Advanced Drug Delivery Reviews, 64*, 1129–1141. https://doi.org/10.1016/j.
addr.2012.04.008

Chapter 10
Design of In Vitro Culture and Bioreactor

Once the tissue-engineered scaffold is constructed, the next job is to seed the cells onto it and culture them in vitro. After that, we can use the completed tissue-engineered device for its final application. While it is theoretically possible to start such a culture with any given number of cells, a minimum starting number of cells usually exist to construct an appropriate tissue-engineered device. The volume of the in vitro culture may also affect the growth of cells. Oxygen diffusion also plays a vital role in tissue development, and as such, the scaffold and in vitro culture must be carefully designed to prevent the cells from suffocating. All these parameters affect the design of in vitro tissue culture, and eventually, the bioreactor design. These parameters can be analyzed by the characteristic times that will be discussed in this chapter. Time constant concepts are well defined in and borrowed from the early tissue engineering textbook – Palsson and Bhatia (2004).

Inquiry 1. Under what conditions do the cells stop proliferating? Have you heard about the Hayflick limit?

Inquiry 2. What are the negative impacts of low oxygen supply? (Refer to Chap. 3.)

10.1 Design of In Vitro Culture

Once the tissue-engineered (TE) scaffold is constructed, cells should be seeded and proliferated on it. The following parameters should be considered:

- The initial number of cells (N_0; the initial cell density $X_0 = N_0/V$ can also be used.)
- The final number of cells (N; the final cell density $X = N/V$ can also be used.)
- The time needed for culture (t)
- The volume of an in vitro culture or a bioreactor (V)
- Oxygen consumption rate (q_{O2}; oxygen uptake rate $= OUR = q_{O2} \times X$ can also be used.)
- Oxygen diffusion length (L_{O2})

© Springer Nature Switzerland AG 2022
J.-Y. Yoon, *Tissue Engineering*, https://doi.org/10.1007/978-3-030-83696-2_10

- Glucose consumption rate (q_{glu})
- Glucose diffusion length (L_{glu}; note: this is less important than L_{O2}.)

We can optimize all these parameters by using various characteristic times. *Characteristic time* t_c is defined as (Palsson and Bhatia, 2004):

$$t_c = \frac{\Delta f}{df \, / \, dt}$$

where f is a function of a specific cell culture parameter.

10.2   Doubling Time (t_d)

By using *doubling time* (t_d), we can calculate the initial number of cells (N_0 or X_0) or the time needed for culture (t) before the in vitro culture. From Chap. 3, we know that cell growth generally follows the first-order growth kinetics, where μ is the specific growth rate (typically in the unit of h^{-1}) (Sect. 3.11):

$$\frac{dX}{dt} = \mu X$$

Solving this simple first-order differential equation with the initial condition of $X = X_0$ at $t = 0$:

$$\ln \frac{X}{X_0} = \mu t$$

For mammalian cell culture, doubling time t_d is usually preferred over specific growth rate μ. *Doubling time* (also known as *turnover time*) t_d is defined as the time for given cells to double their number. Thus $X = 2X_0$ at $t = t_d$.

$$\ln \frac{2X_0}{X_0} = \ln 2 = \mu t_d$$

$$t_d = \frac{\ln 2}{\mu} \text{ or } \mu = \frac{\ln 2}{t_d}$$

Plugging the very first equation (definition of first-order growth kinetics) to this doubling time equation yields:

$$t_d = \frac{\ln(2X_0) - \ln X_0}{\dfrac{1}{X} \dfrac{dX}{dt}}$$

Hence, you can see that doubling time t_d fits the definition of characteristic time where the function f is $\ln X$, since:

$$\frac{d \ln X}{dt} = \frac{1}{X}$$

The final number of cells for various organs is well known for humans. Typical cell density (X) for a human is around 10^8 cells/mL. Assuming a body weight of 70 kg and thus 70,000 mL ($= V$) for a human, there are about 10^{12}–10^{13} cells ($= N$) in one human. Typical organs have volumes from 100 to 500 mL ($= V$), and thus the corresponding cell number in a typical organ is 10^9–10^{11} cells ($= N$). One pancreas has around 10^9 β-islet cells, which produce insulin to regulate blood glucose levels. One liver has a volume of 1–2.5 L and has a total of around 10^{11} cells.

The doubling times of various mammalian cells are summarized in Table 10.1. For most mammalian cells, doubling times are typically 1–3 days. However, for certain types of cells, doubling times are very long, for example, a few hundred days for renal interstitial cells and hepatic cells.

Let us calculate the time to reach 10^9 cells from one cell with the doubling time of (A) 1 day and (B) 3 days. Plugging $\mu = \dfrac{\ln 2}{t_d}$ into $\ln \dfrac{X}{X_0} = \mu t$ gives:

$$\ln \frac{X}{X_0} = \frac{t \ln 2}{t_d} \text{ and thus } t = \frac{t_d \ln(X/X_0)}{\ln 2}$$

With the fixed V, $X/X_0 = N/N_0$.
With $t_d = 1$ day:

$$t = \frac{(1\,day)\ln(10^9/1)}{\ln 2} = 30\,days$$

With $t_d = 3$ days:

$$t = \frac{(3\,days)\ln(10^9/1)}{\ln 2} = 90\,days$$

Table 10.1 Doubling time (t_d) of various mammalian cells

Hematopoietic stem cells	Human	2.5 days
Small intestinal epithelial cells (inner surface of the small intestine)	Human	4–6 days
	Rat	1–2 days
Keratinocytes (outer surface of skin)	Human	2–4 days
Corneal epithelial cells (outermost layer of cornea)	Human	7 days
Lymphatic cells	Rat	7–15 days
Renal interstitial cells	Mouse	165 days
Hepatic cells	Rat	400–500 days

There is an alternative method of calculating the time to reach 10^9 cells from one cell with a given doubling time. With n = number of doublings:

$$X = 2^n X_0 \text{ and } 2^n = \frac{X}{X_0}$$

$$n = \frac{\ln(X / X_0)}{\ln 2}$$

Plugging in $N/N_0 = X/X_0 = 10^9/1$:

$$n = \frac{\ln(10^9 / 1)}{\ln 2} = 30\,\text{doublings}$$

With the doubling time of 1 day, 30 doublings take 30 days. Similarly, with the doubling time of 3 days, 30 doublings take 90 days.

Question 10.1 Calculate the time to reach 10^{11} cells from one stem cell, whose doubling time is 1 day.

Repeating the above example for the hepatic cells with $t_d = 500$ days:
With $t_d = 500$ days:

$$t = \frac{(500\,\text{days})\ln(10^9 / 1)}{\ln 2} = 15000\,\text{days} = 41\,\text{years}$$

We can easily conclude that it is not possible to cultivate such cells in an in vitro culture. There are two possible solutions: (1) to start with a lot higher number of cells or (2) to use stem cells whose doubling time is substantially shorter and differentiate them once a desired number of cells is reached.

Many cells can only divide for a limited number of doublings. This limitation is particularly true for somatic cells (note that the antonym to stem cell is somatic cells, as explained in Chap. 5, as some stem cells are adult cells). Once the somatic cells are differentiated from stem cells, they can usually reproduce for a limited number of doublings, typically from 10 to 50 generations, with an average of 30 generations. This limitation is called the *Hayflick limit*. Somatic cells have *telomeres* at the ends of chromosomes, which are difficult to replicate. As a result, telomeres are shortened after each doubling, associated with this limit in doublings (Fig. 10.1).

Let us assume that the somatic cell line has been isolated from differentiated stem cell culture and has already undergone 10 doublings. Its Hayflick limit is 30 doublings. This means that it can only undergo 20 further doublings. Starting from one cell, it can double for 20 (= 30 − 10) times. We can calculate the maximum number of cells:

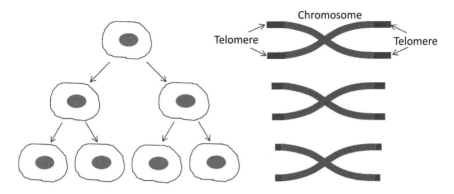

Fig. 10.1 Shortening of telomeres. Length and shortening of telomere are not drawn on a scale

$$X = 2^n X_0 = 2^{20} \times 1 = 10^6 \text{ cells}$$

To reach 10^9 cells, therefore, it is necessary to start with 10^3 cells.

Question 10.2 The somatic cell line has already undergone 15 doublings, and its Hayflick limit is 30 doublings. To reach 10^{11} cells, calculate the minimum starting number of cells. With the doubling time of 3 days, calculate the time to get 10^{11} cells from this minimum number of cells.

Question 10.3 A tissue-engineered transplant is constructed in vitro. Somatic cells are harvested from a donor and have already undergone 20 generations of doublings. These cells have the Hayflick limit of 30 generations (doublings) after freshly differentiated from stem cells. The doubling time for this cell is 36 h. To create a tissue-engineered transplant of 200 mL with a cell density of 10^8 cells/mL, calculate the minimum number of cells that need to be seeded. Using this minimum number, also estimate the time to reach the desired cell density for the given transplant.

10.3   Mean Residence Time (t_{res})

As we have learned from Chap. 3, cell culture media should be replaced periodically, and cells should also be passaged regularly. This feeding and passaging provides fresh media, removes wastes, and prevents contact inhibition. These processes are pretty labor-intensive and inappropriate for industrial operations. A better method is to continuously conduct the in vitro culture, where nutrients and culture media are continually fed, and waste and products are continually removed. In such a case, mean residence time t_{res} becomes quite important in designing such a continuous bioreactor.

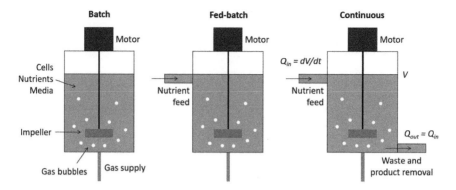

Fig. 10.2 Three types of bioreactor

Bioreactors are classified into three types depending on the feeding and removal conditions (Fig. 10.2):

– In a *batch bioreactor*, nutrients are loaded together with cells and media, and the reactor is sealed. The reactor is stirred with continuous gas (O_2 and CO_2) bubbling to proliferate the cells. Waste and products are continuously accumulated. After a fixed culture time, the reactor is opened, and the cells and products are harvested. The batch bioreactor is not very different from the culture in a tissue culture plate or a microwell plate in a CO_2 incubator, except for the continuous stirring and gas bubbling.
– In a *fed-batch bioreactor*, nutrients are continuously fed into the reactor throughout the cell culture, not to starve the cells. However, wastes and products are not removed during the culture.
– In a *continuous bioreactor*, wastes and products are also continuously removed from the bioreactor. Cells should be retained in the bioreactor, in addition to the continuous nutrient feeding.

All three bioreactors have a motor-driven impeller, a gas supply system, and a temperature control system. Some bioreactors have additional features, including

– A *pH control* system: Normally, the culture medium contains a buffer to maintain an optimal pH. Depending on the waste production, however, a buffer may be insufficient to maintain the optimal pH. For example, lactate wastes decrease the pH as they are acid, while ammonia wastes increase the pH as they are base. In this case, a pH control system may be necessary. A pH electrode is inserted into the bioreactor; If the pH is too high, an acid solution is fed. If the pH is too low, a base solution is provided. Typically, either acid or base solution is needed, but not both, as the pH deviation occurs typically in one direction (i.e., either lactate production or ammonia production).
– *Antifoam* control system: Bioreactors are equipped with an impeller and a motor to provide continuous stirring. While it offers efficient mixing, it sometimes cre-

ates foams. An antifoam agent can be added to the bioreactor depending on the amount of foams in the bioreactor.

- *Dissolved oxygen (DO)* control system: Normally, gas (O_2 and CO_2) is continuously fed into the bioreactor at a fixed rate. For more sophisticated control, a dissolved oxygen (DO) meter can be inserted into the bioreactor and depending on the DO meter reading the gas feeding rate can be controlled.

Mean residence time (t_{res}) is defined as:

$$t_{res} = \frac{V}{Q}$$

where V is the volume (in L), and Q is the volumetric flow rate (in L/h). Note that Q is different from q (metabolic rate). Therefore, t_{res} has the unit of hour (h). It represents the average (mean) time that any single molecule will spend in a given bioreactor. It also fits the definition of characteristic time, where the function f is volume V. Q is $[dV/dt]_{ave}$:

$$t_c = \frac{\Delta f}{df / dt}$$

Mean residence time (t_{res}) can be used to determine Q with the given V, with a specific restriction. For example, we can consider a situation where toxic waste cannot stay in a given bioreactor any more than 6 h. We also want to maximize time for a medium to interact with the cells. In this case, the maximum allowable mean residence time should be 6 h. With the reactor size of 500 mL:

$$Q = \frac{V}{t_{res}} = \frac{500\,mL}{6\,h} = 83\,mL\,/\,h$$

10.4   Oxygen Depletion Time (t_{O2dep})

We must feed cells with sufficient nutrients (e.g., glucose) as well as oxygen. Therefore, we can consider characteristic times for nutrient and oxygen depletion, for example, glucose depletion time ($t_{glu-dep}$) and *oxygen deletion time* (t_{O2dep}). Let us learn about the oxygen depletion time first.

Both glucose depletion and oxygen depletion can be considered chemical reactions. We can define the characteristic times for any given chemical reaction (*characteristic time for reaction*) as:

$$t_{rxn} = \frac{c}{r}$$

where c is the molar concentration of chemicals (e.g., glucose or oxygen) and r is the reaction rate. *Reaction rate r* for reactant's consumption is defined as:

$$r = -\frac{dc}{dt}$$

You can see that this characteristic time also fits its definition, where the function f is molar concentration c.

For the first-order reaction:

$$r = kc$$

where k is the *rate constant*. Plugging this equation into the definition of the characteristic time for reaction yields:

$$t_{rxn} = \frac{c}{kc} = \frac{1}{k}$$

where t_{rxn} is inversely related to the rate constant k.

Oxygen (O_2) has low solubility in water, and the typical amount of *dissolved oxygen (DO)* in tissue is around 0.2 mM. However, most of these DO are consumed at a fast rate and must be replenished through respiration. While it is possible to produce ATPs with very little O_2 (anaerobic condition), ATP production under anaerobic conditions is minimal, as we learned in Chap. 3. The typical oxygen consumption rate in human cells is around 1×10^{-8} mg O_2 / cell-hr. Considering the molecular weight of $O_2 = 32$ g/mol $= 32$ mg/mmol, the oxygen consumption rate q_{O2} is defined as:

$$q_{O2} = 1 \times 10^{-8} \, \frac{mg}{cell \cdot h} \times \frac{1}{32 \, mg \, / \, mmol} = 0.03 \times 10^{-8} \, \frac{mmol}{cell \cdot h}$$

Oxygen uptake rate (OUR) is defined as:

$$OUR = q_{O2} \times X = 0.03 \times 10^{-8} \, \frac{mmol}{cell \cdot h} \times X \, \frac{cell}{mL} = 0.03 \times 10^{-8} \times X \, \frac{M}{h}$$

as mmol/mL = mol/L = M. OUR is the reaction rate for oxygen consumption (= depletion), which is the denominator for the characteristic time for reaction (= r). Plugging in the typical DO = 0.2 mM = 0.2×10^{-3} M in the numerator (= c) yields:

$$t_{O2dep} = \frac{0.2 \times 10^{-3} \, M}{0.03 \times 10^{-8} \times X \, M \, / \, h} = \frac{6.7 \times 10^5}{X} h$$

With X = 10^8 cells/mL:

$$t_{O2dep} = \frac{6.7 \times 10^5}{10^8} = 6.7 \times 10^{-3}\,h\frac{60\,min}{1\,h} = 0.4\,min$$

This result indicates that the cells will be suffocated only after 0.4 min. Note that this is just an example as the level of DO, q_{O2}, and X will vary significantly from tissue to tissue.

Question 10.4 Calculate the oxygen depletion time for a tissue with the "low" cell density $X = 10^7$ cells/mL and the oxygen consumption rate $q_{O2} = 1 \times 10^{-8}$ mg/cell-h. Repeat this question with $X = 10^8$ cells/mL and the "slower" oxygen consumption rate $q_{O2} = 1 \times 10^{-9}$ mg/cell-h. Assume DO = 0.2 mM.

We can repeat the same procedure for glucose consumption rate q_{glu} and the glucose depletion time $t_{glu-dep}$. However, as glucose's solubility to water is a lot higher than that of O_2, the numerator (c) in the $t_{glu-dep}$ is generally far higher than that of t_{O2dep} and the resulting glucose depletion time is almost always much longer than the oxygen depletion time. Therefore, $t_{glu-dep}$ is not considered in designing the in vitro culture, except for extreme, rare cases.

10.5   Oxygen Diffusion Time (t_{O2diff})

Both nutrients (e.g., glucose) and oxygen must diffuse through the tissue to be properly delivered to every cell. We can also define the characteristic times for such *diffusion* of nutrients and oxygen.

Molecules spread in random directions through diffusion. *Fick's first law of diffusion* dictates such molecular diffusion:

$$J = -D\frac{dc}{dx}$$

This equation is the definition for only one direction (x), and 3D diffusion is defined in a more complicated form. J is the *diffusion flux* and has the unit of mass per unit area per time, for example, mol/cm^2-s, and c is the molar concentration (M). D is the diffusivity and has the unit of cm^2/s. Multiplying the diffusivity D (in cm^2/s) by time t (in s) gives the *mean square displacement* (L^2) of molecules. For the molecules diffusing in all three dimensions ($\times 3$) and in forward and backward ($\times 2$):

$$L^2 = 6Dt$$

For the molecules diffusing perpendicular from a blood vessel (Fig. 10.3, left), they can diffuse in two dimensions ($\times 2$) and in forward and backward ($\times 2$):

$$L^2 = 4Dt$$

And for the molecules diffusing from a planar surface (Fig. 10.3, right), they can diffuse only in one dimension ($\times 1$) and in one direction ($\times 1$):

$$L^2 = Dt$$

In general, the diffusion time can be defined as:

$$t_{\text{diff}} = \frac{L^2}{6D}$$

As D is a function of dc/dx, this equation also fits the definition of characteristic time. Diffusivity values of various molecules are:

– Albumin: $D = 6 \times 10^{-7}$ cm²/s
– Glucose: $D = 6 \times 10^{-6}$ cm²/s
– Oxygen: $D = 2 \times 10^{-5}$ cm²/s

With the given diffusion time, it is possible to calculate L (diffusion length). For example, with $t_{\text{diff}} = 2$ min $= 120$ s and glucose $D = 5 \times 10^{-6}$ cm²/s:

$$L = \sqrt{6Dt_{\text{diff}}} = \sqrt{6 \times \left(6 \times 10^{-6} \text{ cm}^2 / \text{s}\right) \times 120 \text{ s}} = 0.066 \text{ cm}$$

indicating that glucose can diffuse 0.066 cm = 0.66 mm = 660 μm for three-dimensional and bidirectional diffusion. For the same with oxygen $D = 2 \times 10^{-5}$ cm²/s:

$$L = \sqrt{6Dt_{\text{diff}}} = \sqrt{6 \times \left(2 \times 10^{-5} \text{ cm}^2 / \text{s}\right) \times 120 \text{ s}} = 0.12 \text{ cm}$$

indicating that oxygen can diffuse 0.12 cm = 1.2 mm = 1200 μm for three-dimensional and bidirectional diffusion.

Likewise, we can also calculate the diffusion time with the given tissue-engineered scaffold design. If the tissue-engineered scaffold was designed with the network of blood vessel supplies that are separated by 1 mm = 0.1 cm in three-dimensional and bidirectional diffusion:

$$L^2 = Dt$$

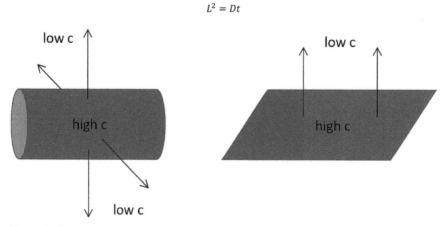

Fig. 10.3 Examples of molecular diffusion

$$t_{\text{diff}} = \frac{L^2}{6D} = \frac{(0.1\,\text{cm})^2}{6 \times 6 \times 10^{-6}\,\text{cm}^2/\text{s}} = 278\,\text{s for glucose}$$

$$t_{\text{diff}} = \frac{L^2}{6D} = \frac{(0.1\,\text{cm})^2}{6 \times 2 \times 10^{-5}\,\text{cm}^2/\text{s}} = 83\,\text{s for oxygen}$$

Question 10.5 Calculate the diffusion times for albumin ($D = 6 \times 10^{-7}$ cm²/s) with the characteristic length $L = 0.1$ cm for (A) three-dimensional, bidirectional diffusion, (B) two-dimensional, bidirectional diffusion, and (C) three-dimensional, one-directional diffusion.

10.6 Design of In Vitro Culture Considering Both Oxygen Depletion Time and Oxygen Diffusion Time

In practical applications, several characteristic times must be considered together for optimum design of in vitro culture.

In Sect. 10.4, we have calculated the oxygen depletion time using the following parameters:

- $q_{O2} = 1 \times 10^{-8}$ mg/cell-h
- DO = 0.2 mM
- $X = 10^8$ cells/mL

And the resulting t_{O2dep} was:

- $t_{O2dep} = 0.4$ min

With this oxygen depletion time, the tissue-engineered scaffold must be constructed to deliver oxygen to every cell in the structure. To make this happen, the oxygen depletion time and oxygen diffusion time must be the same, at the minimum. Assuming three-dimensional and bidirectional diffusion:

$$t_{O2dep} = t_{O2diff} = \frac{L^2}{6D}$$

Plugging $t_{O2dep} = 0.4$ min = 24 s and $D = 2 \times 10^{-5}$ cm²/s for oxygen:

$$L = \sqrt{6 \times 2 \times 10^{-5}\,\text{cm}^2/\text{s} \times 24\,\text{s}} = 0.054\,\text{cm}$$

indicating that the farthest point from each vessel should be not more than 0.054 cm = 0.54 mm = 540 μm, or the vessel-to-vessel distance should be 540 μm × 2 = 1080 μm (Fig. 10.4).

Fig. 10.4 A network of
vessels is designed to make
the oxygen diffusion time
and the oxygen depletion
time the same

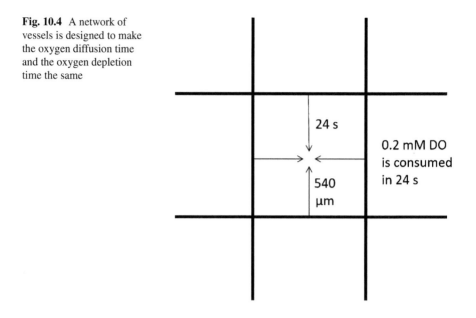

    In general, if the characteristic time for reaction (e.g., oxygen depletion time) is
longer than the characteristic time for diffusion (e.g., oxygen diffusion time), it
takes longer time for the reaction to be completed than for the diffusion to be
achieved.

Question 10.6 Repeat this calculation for glucose, with $c = 2$ mM (higher than
DO), $q_{glu} = 1 \times 10^{-8}$ mg/cell-h (same as q_{O2}), molecular weight of glucose $= 180$ g/
mol, $X = 10^8$ cells/mL (same), and $D = 6 \times 10^{-6}$ cm^2/s (glucose).

Question 10.7 A tissue-engineered transplant is designed with an artificial vasculature. Here are the specifications of this graft:

– Cell density $X = 8 \times 10^7$ cells/mL
– Volume $= 200$ mL
– Oxygen consumption rate $q_{O2} = 10^{-8}$ mg/cell-h
– Dissolved oxygen (DO) in the tissue $= 0.2$ mM
– Diffusivity of oxygen $D = 2 \times 10^{-5}$ cm^2/s

(A) Calculate the oxygen depletion time in the unit of seconds. Be careful with unit
     conversion.
(B) To deliver oxygen from the vasculature to all parts of the graft, the oxygen
     depletion time should be the same as the oxygen diffusion time. From this,
     calculate the characteristic length (L) of the vasculature in the unit of μm.
     Assume three-dimensional, one-directional diffusion.

10.7   Design of Tissue-Engineered Device Using Characteristic Time

All tissue-engineered devices (transplants, organ-on-a-chip, etc.) are mainly constructed from a unit structure, each containing a few thousands of cells. With the cell density of 10^8 cells/mL, 1000 cells correspond to:

$$\frac{1000\,cells}{10^8\,cells\,/\,mL} = 10^{-5}\,mL = 10^{-5}\,cm^3 = 10^7\,\mu m^3$$

This number corresponds to the unit length of:

$$\sqrt[3]{10^7\,\mu m^3} \approx 200\,\mu m$$

This unit length is in the same order as what we have calculated above while slightly smaller than them, indicating that oxygen diffusion will be sufficient for most cases. Natural tissues in human bodies are actually made from a unit structure with a dimension of 100–200 μm. This number is also the typical distances between capillary vessels, alveoli in the lung, and nephrons in the kidney, etc.

Figure 10.5 shows an example of the tissue-engineered scaffold in a mesh form, where the cells are seeded and grown on it. Under a static condition, both nutrients and oxygen are diffused passively. The mesh-form scaffold "obstruct" the natural diffusion, generating many dead cells. Adding an active "flow" to this scaffold can resolve this issue, where the media and oxygen are transported in a forced manner. A more sophisticated approach is introducing a blood vessel network to the TE transplant, which will be discussed in Chap. 13.

10.8   Tissue Engineering Bioreactor

For tissue engineering applications, cells are cultured in two steps:

– Growth bioreactor: Cells are cultured in a conventional manner to increase the cell number, modulate the tissue development, and make the cells function in the desired way. They can be cultured in a tissue culture plate (TCP) in a CO_2 incubator (for a small-scale operation), a batch or fed-batch bioreactor (for a medium-scale operation), and a continuous bioreactor (for a large-scale operation).
– Implantation environment: Cells are then seeded on a scaffold (or an organ-on-a-chip) to achieve focal adhesion, reach confluency, and make them exhibit tissue-like behaviors.

While conventional tank-shaped bioreactors (Fig. 10.2) can be used as growth bioreactors, the bioreactor for the implantation environment may require a specific design due to the scaffold's presence. One such example is a *perfusion bioreactor*,

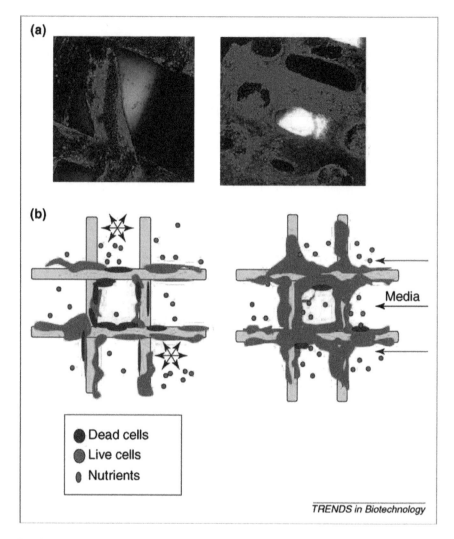

Fig. 10.5 Nutrient transport on a tissue-engineered scaffold (Hutmacher & Singh, 2008. Reprinted with permission. (C) 2008 Elsevier)

shown in Fig. 10.6. The reactor is similar to the tissue culture plate (TCP). The only differences are (1) nutrients and continuously fed in while wastes are pumped out using a syringe pump, and (2) the scaffold is placed within the perfusion bioreactor. There is no motor-driven impeller for mechanical agitation. Gas supply, pH control, antifoam control, DO control, etc., are typically not available, thus inferior to the continuous bioreactor. However, it does provide the continuous supply of nutrients and removal of wastes, not found in typical in vitro TCP culture.

It is also possible to use multiple scaffolds to mimic the repeating units in an organ, for example, nephrons in a kidney, lobules in a liver, etc. Figure 10.7 shows

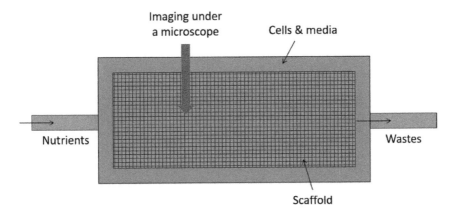

Fig. 10.6 Perfusion bioreactor

Fig. 10.7 Perfused
column bioreactor

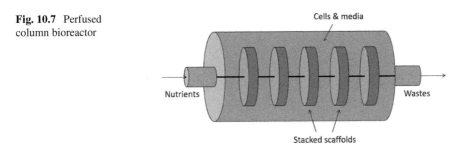

the *perfused column bioreactor*, where multiple scaffolds are used in a cylindrical-shaped bioreactor. (Of course, a single scaffold can be used in a perfused column bioreactor, although it undermines this bioreactor's benefit.) It may be possible to rotate the entire cylinder to provide mechanical agitation to the bioreactor.

A *hollow fiber bioreactor* is another good example (Fig. 10.8). A hollow fiber module is a bundle of hollow fibers, which are tiny tubes that are semipermeable. While it has prominently been used for water purification, it can also be used as a bioreactor for tissue engineering. Media and nutrients are fed into the hollow fibers. They can either pass through the other end of hollow fibers and recycled back to the inlet or permeate through the hollow fibers. Cells are anchored to the exterior surfaces of hollow fibers and uptake the media and nutrients permeated from the inside of the hollow fibers. Cells are too big to permeate through the hollow fibers (e.g., semipermeable). There is a secondary inlet on the module's shell side, where we can continuously feed growth factors, differentiation factors, etc. They travel primarily outside the hollow fibers, where the cells are found. As more sophisticated delivery of nutrients and other factors is possible, hollow fiber bioreactors have popularly been used as a growth bioreactor, especially when sophisticated differentiation and tissue development are necessary. The hollow fiber bioreactor can sometimes be used as an implantation environment as its structure is similar to specific organs, for example, the kidney.

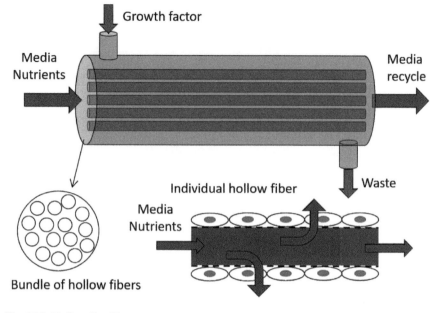

Fig. 10.8 Hollow fiber bioreactor

10.9   Measurements from a Tissue Engineering Bioreactor

How can we make sure the cells function in the desired manner and the tissue-engineered transplant or organ-on-a-chip exhibit tissue-like behaviors? We can measure the followings from the bioreactor:

– Growth rates (μ or t_d)
– Metabolic rate ($q_{glucose}$, $q_{glutamine}$, $q_{lactate}$, $q_{ammonia}$, q_{O2}, etc.)
– Extent of focal adhesion
– Protein expression on the cell membrane (integrins, cadherins, etc.)
– Protein production (e.g., albumin from hepatocytes)
– ECM deposition (e.g., collagen fibers)
– Mechanical properties of the tissue

10.10   Laboratory Task 1: Design of In Vitro Culture and Co-culture

Instead of hands-on experiments, we will conduct a series of in vitro culture engineering designs for tissue engineering applications.

Objective 1. Design of In Vitro Culture

Consider a tissue-engineered transplant with a total volume of 1 L. This transplant's specific cells can be proliferated only up to 10^3 fold in an in vitro culture. The doubling time for this cell is 48 h. The final cell density should be 10^8 cells/mL.

(A) What is the minimal number of cells required initially?
(B) Using this minimal number, calculate the amount of time in hours to reach the final cell density.

Objective 2. Design of In Vitro Co-culture

A tissue-engineered skin transplant is being developed in an in vitro culture using an appropriate polymeric scaffold. Two different cell types are being co-cultured in a bioreactor:

	Doubling time (t_d)	Number of cells in the skin (%)
Keratinocytes (makes up epidermis)	66 h	60
Fibroblasts (makes up dermis)	21 h	40

The bioreactor will be operated for 20 days, by when the overall cell density becomes 5×10^7 cells/mL with the total volume of 100 mL (not counting the volume of a polymeric scaffold). Calculate the number of keratinocytes and fibroblasts that need to be initially seeded onto the polymeric scaffold. Both cells follow the first-order growth kinetics. Hint: Use the absolute cell number, not cell density.

10.11 Laboratory Task 2: Design of Artificial Vasculature Using Oxygen Depletion and Diffusion Times

A tissue-engineered transplant has been designed with an artificial vasculature. Oxygen and glucose must be able to be delivered from the vasculature to all parts of the graft. The following properties are provided for a completed, tissue-engineered graft:

- Cell density $X = 5 \times 10^7$ cells/mL
- Oxygen consumption rate $q_{O2} = 10^{-8}$ mg/cell-h
- Rate constant of glucose consumption reaction $k = 10^{-3}$ s^{-1}
- Dissolved oxygen (DO) in the tissue $= 0.2$ mM
- Diffusivity of oxygen $D = 2 \times 10^{-5}$ cm^2/s
- Diffusivity of glucose $D = 6 \times 10^{-6}$ cm^2/s

(A) Calculate the oxygen depletion time in seconds. Make sure to convert mol to g unit.
(B) Using this time (t_{O2dep}), calculate the characteristic length L (the distance of the most distant interior point from the surface) of the vasculature in mm. Assume three-dimensional, bidirectional diffusion.

(C) Calculate the glucose depletion time in seconds. Assume the glucose consumption reaction follows the first-order reaction kinetics. In such a case:

$$t_{rxn} = \frac{c}{r} = \frac{c}{kc} = \frac{1}{k}$$

(D) Using this time ($t_{glu-dep}$), calculate the characteristic length L of the vasculature in mm. Again, assume three-dimensional, bidirectional diffusion. Between oxygen depletion and glucose consumption, which one requires denser vasculature?

Review Questions
1. Define and explain various characteristic times: (1) doubling time, (2) mean residence time, (3) characteristic time for reaction, (4) oxygen depletion time, (5) glucose depletion time, (6) characteristic time for diffusion, (7) oxygen diffusion time, and (8) glucose diffusion time.
2. Calculate the initial number of cells that are needed to build a tissue-engineered device with given restrictions. Calculate the time required to obtain the necessary number of cells for a given tissue-engineered device. Repeat the question with a co-culture scenario.
3. Repeat 2 with the limitation of the Hayflick limit.
4. Compare batch, fed-batch, and continuous bioreactors. Discuss their strengths and weaknesses in tissue engineering applications.
5. Compare perfusion, perfused cylinder, and hollow fiber bioreactors. Discuss their strengths and weaknesses in tissue engineering applications.
6. Calculate the characteristic diffusional length of oxygen, glucose, a small protein, a large protein, etc., with given parameters.

References

Hutmacher, D. W., & Singh, H. (2008). Computational fluid dynamics for improved bioreactor design and 3D culture. *Trends in Biotechnology, 26*, 166–172. https://doi.org/10.1016/j.tibtech.2007.11.012
Palsson BO, Bhatia SN. (2004). *Tissue engineering.* Pearson Prentice Hall. ISBN 978-0130416964.

Chapter 11
Organ-on-a-Chip

Starting from this chapter, we will cover the practical applications of tissue engineering. The first example is organ-on-a-chip (OOC). OOC mimics the human organ, which can be used for many different applications, including drug tests (efficacy and toxicity), disease models, etc. While the final OOC is not transplanted back to humans, it is made from tissue engineering principles and technologies that we have learned throughout this book. (It is indeed possible to transplant the OOC to a human; however, it will then become a tissue-engineered transplant rather than being an OOC.)

Inquiry 1. Drug tests and disease models have utilized (1) in vitro cell culture, (2) animal tests, and (3) human clinical trials. In your own words, identify potential problems with this approach.

Inquiry 2. In your own words, how can OOC be positioned to the above procedure (cell culture–animal test–clinical trial)? Can OOC replace one or two of this procedure?

11.1   2D Versus 3D Cell Culture

In the earlier chapters, we have learned cell culture mainly in 2D. Anchorage-dependent cells are cultured in a petri dish, a T-25 tissue culture plate (TCP), or a microwell plate. Microwell plate helps conducting many cell cultures under varying conditions (e.g., up to $96 = 8 \times 12$) of cell cultures. These cell cultures can be used for evaluating the efficacy and toxicity of drugs (including vaccines), as explained in Chap. 2.

However, 2D cell culture cannot mimic cell–cell and cell–ECM interactions correctly. As a solution, *3D cell culture* has been proposed. The anchorage-dependent cells are grown in a 3D hydrogel matrix, where the cells can proliferate and interact

© Springer Nature Switzerland AG 2022
J.-Y. Yoon, *Tissue Engineering*, https://doi.org/10.1007/978-3-030-83696-2_11

2D cell culture   3D cell culture

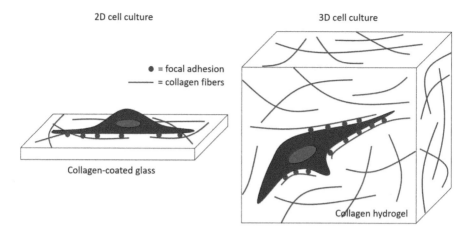

● = focal adhesion
—— = collagen fibers

Collagen-coated glass

Collagen hydrogel

Fig. 11.1 (Identical to Fig. 6.2) Cells on a 2D surface (collagen-coated glass) and a 3D gel (collagen gel)

with the surroundings in all three dimensions. 3D cell culture has already been explained in Chap. 6 (Sect. 6.2 and Fig. 6.2).

Unfortunately, 3D cell culture is still not satisfactory in correctly mimicking the tissue. There is no interface between different tissues, for example, *epithelium*. It is also challenging to recapitulate the chemical (e.g., growth factors), nutrient, and oxygen gradients occurring in the native tissue. Most importantly, there is no active liquid flow in the 3D cell culture. Most tissues comprise tubular structures where the liquids are transported actively, for example, sinusoids in the liver, nephrons in the kidney, etc. This lack of flow also leads to the absence of mechanical stimuli (e.g., shear, compression, etc.), critical in maintaining the somatic cells' correct metabolism and stem cells' differentiation (Fig. 11.1).

11.2   Organ-on-a-Chip

Organ-on-a-chip (OOC) has evolved from *lab-on-a-chip* (*LOC*). LOC is a network of channels and wells fabricated on a silicone-based substrate using photolithography or soft lithography technique. Liquids flow through these channels and wells, being mixed, incubated, and separated in sub-millimeter (= micrometer) scale channels. Such liquid manipulation is called *microfluidics*, and LOC is sometimes referred to as *microfluidic devices* or *microfluidic chips*. However, the exact definitions of LOC and microfluidic device/chip may slightly differ from each other. These processes occur in a miniaturized manner, saving the sample and reagent volume and potentially allowing for field-based applications. The number of channels can easily be multiplied, allowing high-throughput processing. LOC has been used for various applications, including high-throughput separation and purification, drug development, chemical analysis, etc. The most popular LOC application is the point-of-care diagnostics (POCT), where the user conducts the traditional laboratory-based diagnostics (immunoassays, polymerase chain reaction, etc.) in a miniaturized and rapid manner.

As the LOC-based POCT has become mature, new attempts to manipulate or even proliferate mammalian cells in a LOC have been demonstrated. Eventually, such attempts have been converged to proliferate the mammalian cells in a LOC to create a tissue or organ mimic. *Organ-on-a-chip* (*OOC*) is a 3D cell culture enhanced with the LOC (or microfluidics) technology. OOC can provide the following features in addition to the 3D cell culture:

– Precisely tuned dynamic flow
– Providing chemical gradients (growth factors, differentiation factors, hormones, enzymes, etc.)
– Delivering nutrients and oxygen in a controlled manner
– Providing mechanical stimuli (compression, tension, shear, etc.)

Question 11.1 What is the prominent distinguishing feature of organ-on-a-chip from 3D cell culture?

11.3   How Do You Fabricate Lab-on-a-Chip (LOC)?

As briefly explained in the previous section, photolithography and soft lithography are the two most popular methods of fabricating LOCs. *Lithography* is a printing method that has been popularly used for a very long time. In conventional lithography, images or texts are drawn (graph = to write) on a limestone (litho = stone) using oil. Using the immiscibility between oil and water (= ink solution), the images or texts are transferred to paper. In *photolithography* (Fig. 11.2), the limestone is replaced with the silicon wafer or the silicone-based polymer substrate. (Silicon refers to the natural material, while silicone to the synthetic materials.) For both semiconductors and LOCs, the final products are pretty small, flat pieces, or "chips." Therefore, they are called semiconductor chips or lab-on-a-chip. This wafer or substrate is typically spin-coated (refer to Sect. 6.11 and Fig. 6.12) with a *photoresist* (*PR*) material. An appropriate pattern (e.g., an integrated circuit for constructing a semiconductor chip or a microfluidic channel layout for a LOC) is "printed" on a transparent mask. A projection system is used to make the pattern smaller on a micrometer scale. This printed mask is aligned on a silicon wafer (or a silicone-based substrate), and high-energy light (typically UV light) is irradiated on them. The PR exposed to UV can be either fragmented (positive PR) or hardened (negative PR), while the PR not exposed to UV is unaffected. A developer is added to dissolve and remove the fragmented (positive) PR or dissolve and remove the untouched area while retaining the hardened (negative) PR. In this manner, we can transfer the pattern to the semiconductor chip or the LOC.

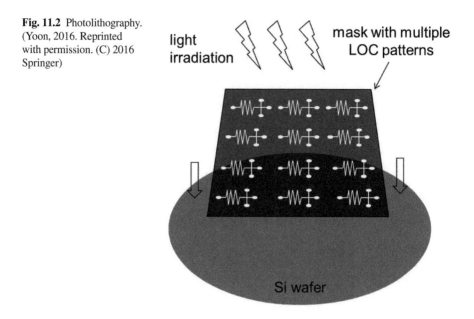

Fig. 11.2 Photolithography. (Yoon, 2016. Reprinted with permission. (C) 2016 Springer)

Figure 11.3 illustrates the overall process of creating a LOC. A "trench" channel is created after the photolithography. A coverslip should be bonded to make this open trench into a closed microchannel. An inlet and an outlet hole should also be punched. These holes introduce the samples and reagents to the LOC and dispense the solution out.

This photolithography process requires a unique laboratory environment (called *clean lab*) and special equipment. *Soft lithography* has been developed as a simpler alternative to photolithography. One of the most popular methods of soft lithography is *PDMS replica molding*, as illustrated in Fig. 11.4. Once a pattern is developed on a silicon wafer (or a silicone-based substrate), we can use it as a mold to create multiple copies of it. Typically, *polydimethylsiloxane (PDMS*; refer to Sect. 6.11 and Fig. 6.14) is used for replica molding. PDMS is also a silicone-based polymer. PDMS comes in the form of a gel, and you can easily pour it on the mold. Once PDMS is settled on the mold, they are cured or cross-linked. You can then peel off the PDMS from the mold, creating one copy. You can repeat the procedure multiple times until the mold becomes unusable. You can perform these procedures in any laboratory without the need for special equipment or a high level of training. The photolithographed mold, however, must be fabricated in a complicated manner. Recently, however, there are many commercial vendors who can create the mold from your design at a reasonable price (a few hundreds of US dollars). Once you secure such a mold, you can make tens and even hundreds (if you are careful in handling the mold) of copies from the same mold. There are other soft lithography techniques available, while the basic concept of creating multiple copies from the same mold remains the same.

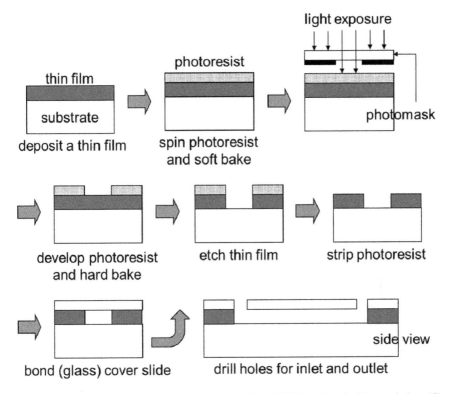

Fig. 11.3 LOC fabrication with photolithography. (Yoon, 2016. Reprinted with permission. (C) 2016 Springer)

PDMS-based soft lithography has popularly been utilized for many OOC applications as there are several benefits over photolithography:

– PDMS is optically transparent while silicon wafer is not. Optical transparency is quite essential in imaging cells (as we have learned throughout this book).
– PDMS is gas-permeable while silicon wafer is not. This permeability is an essential character for properly delivering oxygen to the cells in an OOC.
– PDMS is flexible while silicon wafer is stiff. It is easy to apply compression, tension, and shear to the PDMS-based OOC, which can mimic the in vivo tissues' mechanical stimuli.

Bear in mind that PDMS is strongly hydrophobic, as we have learned in Chap. 6. Such strong hydrophobicity provides an excellent nonstick property for liquid flow and subsequently useful for cardiovascular applications. However, cells will not adhere to the OOC surface well. Typically, the areas where cells are anchored should be coated with ECM proteins, for example, collagens, fibronectin, vitronectin, laminin, etc., or sometimes with blood clots (fibrins) that also exhibit suitable cell adhesion property.

Replica molding

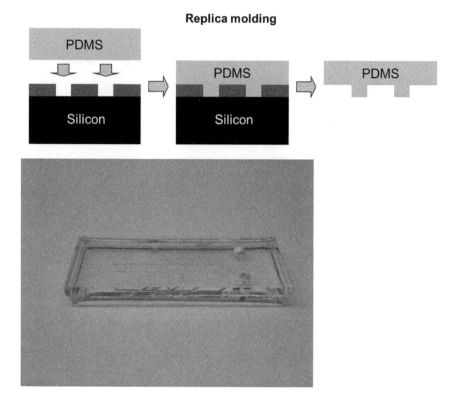

Fig. 11.4 Top: replica molding process; bottom: a LOC made from PDMS replica molding. (Yoon, 2016. Reprinted with permission. (C) 2016 Springer)

11.4   OOC Example: Kidney-on-a-Chip

A *kidney* is an essential organ whose primary function is the filtration of blood and the excretion of wastes into the urine. A kidney is made from a large number of tubular unit structures called *nephrons* (Fig. 11.5). Blood is collected from glomerular capillaries into the Bowman's capsule and enters each nephron unit. The nephron's tubular structure is made from a basement membrane, and the *kidney epithelial cells* are anchored on it (Fig. 11.6). Kidney epithelial cells express a unique, brush-like structure on their surface, called *brush border*. They can selectively reabsorb nonwaste materials and deliver outside the nephron, which are eventually collected back to the capillary blood vessels.

A typical kidney-on-a-chip mimics one nephron made from a single microfluidic channel. In the middle of the microfluidic channel, a porous membrane can mimic the basement membrane. Two identical trench channels are typically fabricated,

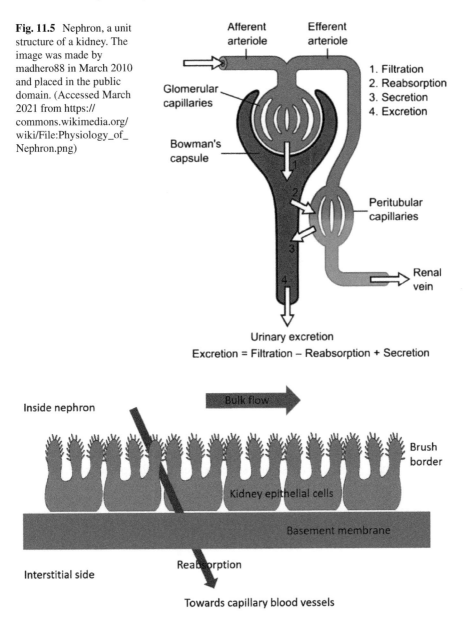

Fig. 11.5 Nephron, a unit structure of a kidney. The image was made by madhero88 in March 2010 and placed in the public domain. (Accessed March 2021 from https://commons.wikimedia.org/wiki/File:Physiology_of_Nephron.png)

Fig. 11.6 Kidney epithelial cells in a nephron

followed by sandwiching a porous membrane in between (Fig. 11.7). An active flow is applied to the top layer, mimicking the flow through a nephron. There is no flow in the bottom layer, mimicking the interstitial layer where the reabsorption back to the capillary occurs eventually.

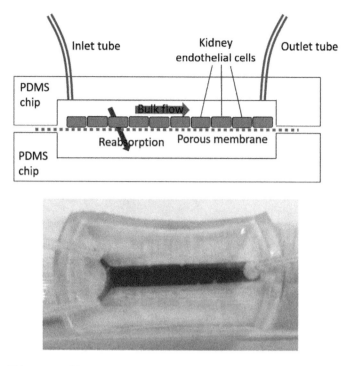

Fig. 11.7 Kidney-on-a-chip

The small intestine's structure is very similar to the kidney's. The epithelial cells inside the small intestine show similar structures, called *villi* (plural form; singular is *villus*), where absorption occurs. Therefore, the resulting intestine-on-a-chip is very similar in its design, while its dimension is substantially bigger than that of the kidney's.

11.5   OOC Example: Liver-on-a-Chip

The liver is a versatile organ with multiple roles, including protein production (e.g., albumin), breakdown of nutrients and toxicants, etc. The liver is made from a large number of hexagonal unit structures called *liver lobules* or *hepatic lobules* (hepar or hepa = liver) (Fig. 11.8). At the outside of each lobule, there are portal veins where the blood is supplied. At the center, there is a single central vein (or *hepatic vein*).

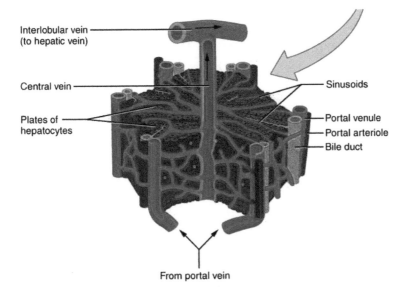

Interlobular vein
(to hepatic vein)

Central vein

Plates of
hepatocytes

Sinusoids

Portal venule
Portal arteriole
Bile duct

From portal vein

Fig. 11.8 Liver lobule, a unit structure of a liver. (Openstax, 2013. (C) 2013 Rice University. Open access article distributed under the terms of the Creative Commons Attribution 4.0 International License)

There are multiple tubular structures from the portal vein to the central vein, called *liver sinusoids* or simply *sinusoids* (Fig. 11.9). The inside of the sinusoid is made from *sinusoidal endothelial cells*, where the outside is covered by cuboidal-shaped *hepatocytes* (hepa = liver; cyte = cell). You can also notice *Kupffer cells*, which fight off the pathogens (*phagocytosis*) similar to the macrophages found in tissues.

Like the nephrons in a kidney, molecules (nutrients, toxicants, proteins, etc.) are transported from inside the sinusoid to the hepatocytes or vice versa to accomplish the nutrient breakdown, toxicant breakdown, and protein production.

A typical liver-on-a-chip mimics one sinusoid made from a single microfluidic channel. Alternatively, we can construct a bundle of sinusoid-like microfluidic channels to mimic a lobule-like structure, as shown in Fig. 11.10. Hepatocytes are patterned on the microchannel structures. The channel is made from a microfabricated surface (lacking sinusoidal endothelial cells), mimicking sinusoidal endothelium's membrane-like nature. In the device shown in Fig. 11.10, paper (nitrocellulose fibers) is used as the membrane-like endothelium. As explained in Chap. 10 (Sect. 10.2 and Table 10.1), hepatocytes' doubling time is exceptionally long, that is, 400–500 days. Therefore, it is not practical to use somatic hepatocytes to build a liver-on-a-chip. *HepG2 cells* are used for many liver-on-a-chip applications. They are immortalized cells and thus proliferate way faster than somatic hepatocytes, with a doubling time of 1–2 days. Despite being immortalized cells, they still show the somatic hepatocytes' characteristics. They metabolize the nutrients and

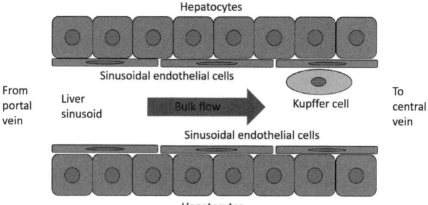

Fig. 11.9 Liver sinusoid

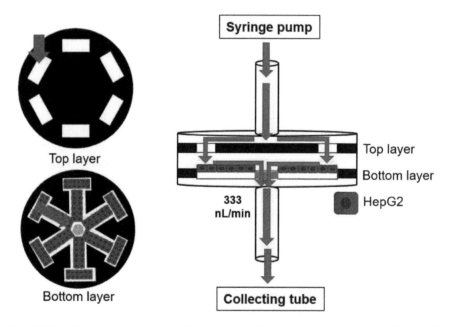

Fig. 11.10 A liver-on-a-chip example. The six outside inlets represent portal veins, and the central outlet represents a central vein, thus creating a liver lobule-like structure. The immortalized cell line of HepG2 (doubling time = 1–2 days) is used instead of somatic hepatocytes (doubling time = 400–500 days). Nitrocellulose paper is used as a substrate to mimic the sinusoidal endothelium (sinusoidal endothelial cells are not used). (Kaarj et al., 2020b. Reprinted with permission. (C) 2020 Springer)

toxicants, and produce proteins. However, HepG2 cells are not fully cuboidal and do not form tight monolayers, unlike the somatic hepatocytes.

11.6   OOC Example: Lung-on-a-Chip

The lung is a vital organ where the gas exchange is made. O_2 is taken from the inhaled air to the blood, while CO_2 is dumped from the blood to the exhaled air. The lung's unit structure is called *pulmonary alveolus* or simply *alveolus* (plural form is *alveoli*), where the air is brought in (inhalation) or squashed out (exhalation) (Fig. 11.11). There are numerous capillary blood vessels in the alveoli where the gas exchange occurs between the inhaled air and the blood.

Lung-on-a-chip is structurally similar to the kidney-on-a-chip. Two channels are sandwiching a porous membrane, where the *alveolar cells* are anchored on the top and the *vascular endothelial cells* to the other side. There is an airflow through the top and blood flow through the bottom side (Fig. 11.12).

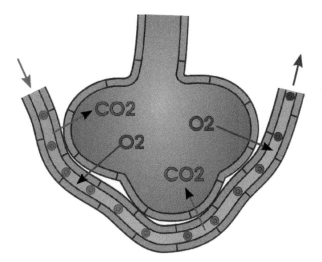

Fig. 11.11 A schematic illustration of the pulmonary alveolus. A capillary blood vessel is shown at the bottom, whose inner surface is covered with vascular endothelial cells. The top cavity is lined with alveolar cells (there are two types – I and II). There are also *alveolar macrophages* (not shown in the figure), which do phagocytosis, similar to the liver's Kupffer cells. The image was made by domdomegg in January 2016 and placed in the public domain. (Accessed March 2021 from https://commons.wikimedia.org/wiki/File:Gas_exchange_in_the_aveolus.svg)

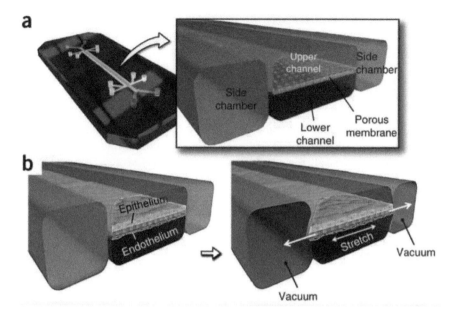

Fig. 11.12 Lung-on-a-chip. (Huh et al., 2013. Reprinted with permission. (C) 2013 Springer Nature)

11.7   OOC Example: Angiogenesis-on-a-Chip

Angiogenesis is the formation of new blood vessels (Fig. 11.13). It constantly occurs throughout the body. Angiogenesis is essential for tissue-engineered transplants as the cells must be fed with nutrients and oxygen in vivo. It can happen naturally in the body after the tissue-engineered device is transplanted, often promoted by adding a specific growth factor for blood vessels, for example, *vascular endothelial growth factor* (*VEGF*). Significant elevation in angiogenesis usually indicates the progression of the malicious tumor, that is, cancer.

We can also mimic angiogenesis on an OOC. Unlike the other OOCs, you should provide an open space where the blood vessels are free to sprout and grow. The inner side of a blood vessel is covered by *vascular endothelial cells*. For OOC applications, *human umbilical vein endothelial cells* (*HUVECs*) are most frequently used. HUVECs are obtained from the human umbilical cord, which is typically discarded after birth. HUVECs are easier to maintain and proliferate quite well compared to the somatic vascular endothelial cells. Figure 11.14 shows the example of angiogenesis-on-a-chip. Two channels are filled with vascular endothelial cells (or HUVECs), mimicking two existing blood vessels. In between these two channels, chemical cues (VEGF) or mechanical cues (compression and shear) are applied. In response to these cues, vessel sprouting can be observed.

Fig. 11.13 Angiogenesis. A new blood vessel "sprouts" from the existing vessel and grows to the other vessel, eventually creating a network-like connection

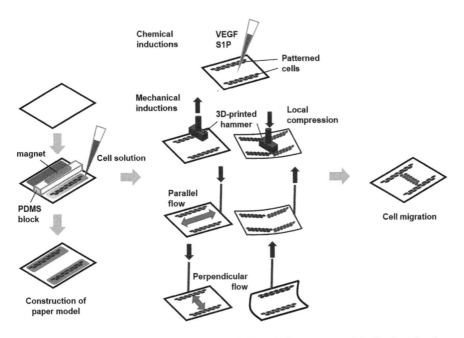

Fig. 11.14 Angiogenesis-on-a-chip. (Kaarj et al., 2020a. (C) Open access article distributed under the terms of the Creative Commons Attribution 4.0 International License)

11.8 OOC Example: Blood–Brain Barrier (BBB)-on-a-Chip

There are numerous blood vessels in the brain, like many other organs in a body. However, the brain is protected from harmful toxicants and disease-associated molecules, called the *blood–brain barrier* (*BBB*). Unfortunately, this BBB also blocks the delivery of many drugs to the brain. Therefore, BBB-on-a-chip has been developed to investigate this BBB and find a way to deliver drugs to the brain effectively.

Figure 11.15 shows the schematic illustration on BBB. It consists of *brain microvascular endothelial cells* (*BMEC*) that make up the brain's blood vessels,

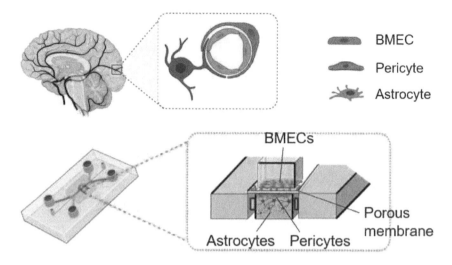

Fig. 11.15 BBB-on-a-chip. (Liang & Yoon, 2021. (C) Open access article distributed under the terms of the Creative Commons Attribution 4.0 International License)

surrounded by pericytes and astrocytes. BBB-on-a-chip is constructed in the same manner as kidney-on-a-chip, where a microfluidic channel is separated by a porous membrane, where the BMECs are anchored on one side and the pericytes and astrocytes on the other side.

11.9   Other OOC Examples

Many other organs and tissues have been mimicked with OOCs. In addition to the kidney, intestine (briefly mentioned along with kidney-on-a-chip), lung, and blood vessel (angiogenesis-on-a-chip), the following organs have also been demonstrated: spleen, breast, skin, muscle, bone, eye (cornea), brain (especially *blood–brain barrier*), etc.

11.10   Multiple-Organs-on-a-Chip and Human-on-a-Chip

Rather than mimicking one organ at a time, two or more organs can be connected to investigate the systemic effect of organ-to-organ communications. Such OOCs are called *multiple-organs-on-a-chip* (*MOCs*). It is also possible to connect not just two or more but a large number of organs together, and it is specifically referred to as *human-on-a-chip* (Fig. 11.16).

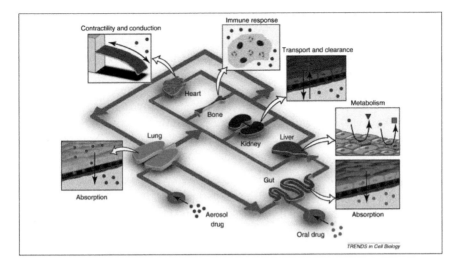

Fig. 11.16   Human-on-a-chip. (Huh et al., 2011. Reprinted with permission. (C) 2011 Elsevier)

11.11   OOC Application: Drug Testing

There are numerous applications of OOCs. The most promising OOC application in terms of commercial viability would be drug testing, specifically evaluating the efficacy and toxicity of drugs. Traditionally, drugs have been tested with (1) *in vitro test*, (2) *animal test,* and (3) *human clinical trial*. In vitro drug tests are usually made on a 2D cell culture, although 3D cell culture has gained popularity. They are generally conducted in a microwell plate (e.g., 96-well). Varying doses of a drug are added to each well, and the cells' viability, morphology, and metabolic status are assessed. On the other hand, OOCs mimic the tissue- and organ-level responses to these drugs, incorporating the ECM components, active flows, and introducing chemical and mechanical cues in a programmed manner. Here are the OOC's advantages over in vitro test:

– Improved recapitulation of an organ with the ECM-like structure (porous membrane, interstitial area, etc.)
– Active flow (blood flow, liquid flow, air flow, etc.)
– Ability to introduce chemical cues growth factors, differentiation factors, etc.) in a programmed manner
– Ability to introduce mechanical cues (compression, tension, shear, etc.) in a programmed manner
– Better control for pharmacokinetic study

Here are the OOC's advantages over the animal test:

– Lower cost than using animals
– Animal models often fail to predict the human response (species difference)

– Ethical issues of sacrificing animals

Toxicity evaluation of environmental toxicants is also investigated in the same manner as drug tests. Likewise, OOCs can also be used to evaluate the toxicity of environmental toxicants (Cho & Yoon, 2017; Akarapipad et al., 2021). The same goes for the toxicity evaluation of cosmetic products.

In drug testing, pharmacokinetics (PK) and pharmacodynamics (PD) are pretty commonly assessed. In *pharmacokinetics* (*PK*), the drug amount residing in a body is monitored over time. *ADME analysis* is the most well-known example of PK, where ADME stands for absorption, distribution, metabolism, and excretion.

– *Absorption* represents the uptake of drugs (or toxicants, cosmetics, etc.) into a human body through the mouth, nose, skin, lung, etc. Intestine-on-a-chip, lung-on-a-chip, and skin-on-a-chip are good models for assessing the drug absorption.
– *Distribution* represents the movement throughout a human body, generally through blood vessels or lymph. Blood vessel-on-a-chip is an excellent model for assessing the drug distribution. Multiple-organs-on-a-chip and human-on-a-chip are also good models for investigating the organ-to-organ distribution of drugs.
– *Metabolism* represents the breakdown of nutrients, toxicants, etc., and the production of proteins. A good example is liver-on-a-chip, where all the above take place in a single organ.
– *Excretion* represents the elimination of wastes to reduce toxic effects. A good example is kidney-on-a-chip.

Sometimes, drugs (or toxicants, cosmetics, etc.) can be accumulated in a specific organ, called *bioaccumulation.* Such bioaccumulation can occur in the liver, kidney, brain, muscle, fat, etc. We can utilize an individual OOC model to study such bioaccumulation (Akarapipad et al., 2021). Multiple-organs-on-a-chip and human-on-a-chip are also excellent in investigating this bioaccumulation.

In *pharmacodynamics* (*PD*), the drug's pharmacological effects are monitored over time. In other words, PD monitors how the organ responds to the drug, whereas PK monitors the drug amount upon ADME. If a drug is designed to reduce or enhance a specific protein's production, we should monitor the amount of such protein production. If a drug is designed to neutralize a particular chemical or electrical signal, the extent of such a chemical or signal should be monitored, etc.

11.12   OOC Application: Disease Model

OOCs can also be used to study various diseases to determine an appropriate drug or treatment option. One of the popular OOC disease models is a cancer model. *Carcinogenesis* (carcino = cancer; genesis = creation) has popularly been evaluated on OOC models. Certain materials in food and water have been suspected to be carcinogenic (cancer-causing). We can expose these materials (typically in a higher concentration for a shorter time duration) to the OOC and monitor the cells' morphology, proliferation, and metabolism and determine their carcinogenicity.

Metastasis has also been evaluated on OOC models, where metastasis indicates the spread of cancer cells to the other parts of a body. Most cancer patients enter their terminal stages because of metastasis. Multiple-organs-on-a-chip and human-on-a-chip are excellent examples of studying metastasis.

We can also investigate stem cell differentiation with OOC models. Morphogenetic factors are responsible for stem cell differentiation. These include growth factors, differentiation factors, ECM shape/structure, and mechanical cues. While recapitulation of ECM shape/structure and providing mechanical cues are problematic in the in vitro cell culture, they can easily be demonstrated in OOCs. We can provide chemical cues (growth factors, differentiation factors, etc.) in a programmed manner.

Question 11.2 From the following list, choose one application that can be benefited the most from human-on-a-chip. Briefly explain why.

A. Stem cell differentiation
B. Metastasis
C. Cosmetics toxicity
D. Biocompatibility of implants

11.13   Mechanical Stimuli to OOCs

Many different chemical cues (growth factors, differentiation factors, hormones, vitamins, nutrient deficiencies, oxygen deficiencies, etc.) have been applied to the OOCs. Mechanical stimuli can also affect organ behavior and early organ development (e.g., stem cells) and have been used to the OOCs. Figure 11.17 summarizes the mechanical stimuli that have been applied to the OOCs. The top channel represents the active fluid flow (e.g., blood vessels, lymphatic vessels, nephrons in a kidney, liver sinusoids, etc.), while the bottom area represents the neighboring interstitial tissue. Programmed flows are typically introduced using a syringe pump (demonstrated in Laboratory Task 2 of this chapter), while compression/stretch/strain is applied using a motor with pressurized air or vacuum.

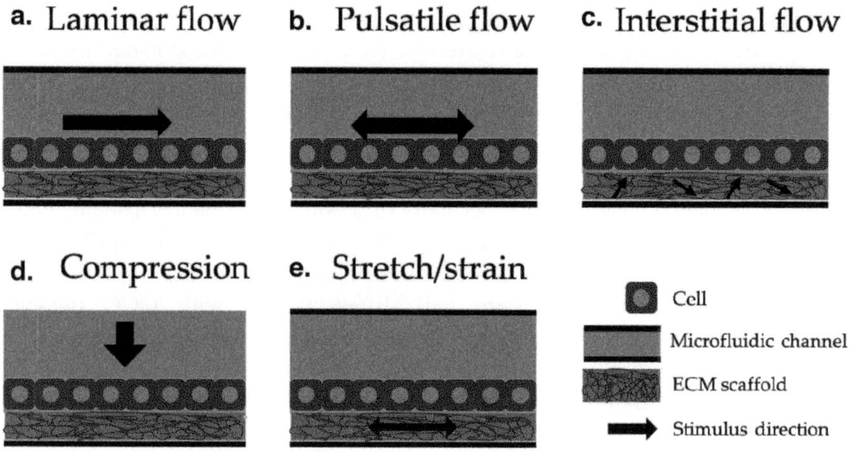

a. Laminar flow   **b.** Pulsatile flow   **c.** Interstitial flow

d. Compression   **e.** Stretch/strain

Cell
Microfluidic channel
ECM scaffold
Stimulus direction

Fig. 11.17 Various mechanical stimuli were applied to the OOCs. (Kaarj & Yoon, 2019. (C) Open access article distributed under the terms of the Creative Commons Attribution 4.0 International License)

11.14   Laboratory Task 1: OOC Fabrication

Laboratory tasks of this chapter are based on two works:

- Cho S, Islas-Robles A, Nicolini AM, Monks TJ, Yoon JY. 2016. In situ, dual-mode monitoring of organ-on-a-chip with smartphone-based fluorescence microscope. *Biosensors and Bioelectronics* 86: 697–705. https://doi.org/10.1016/j.bios.2016.07.015
- Kaarj K, Ngo J, Loera C, Akarapipad P, Cho S, Yoon JY. 2020. Simple paper-based liver cell model for drug screening. *BioChip Journal* 14: 218–229. https://doi.org/10.1007/s13206-020-4211-6

Objective 1. OOC Fabrication by Replica Molding
In this task, OOC with a Y-shaped microfluidic channel, that is, two inlets and one outlet, will be fabricated using the replica molding technique.

1. Design your Y-shaped channel, shown in Figs. 11.4 and 11.7, using appropriate design software (e.g., SolidWorks). Set your channel depth and length somewhere between 100 μm (= 0.1 mm) and 2000 μm (= 2 mm). You can use a commercial service to create the mold fabricated on a silicon wafer. Alternatively, you can print your pattern using a 3D printer, as shown below (Fig. 11.18). We will use the 3D printing method in this laboratory task.
2. Using an adhesive, attach the 3D printed mold inside a petri dish. Mold should be positioned at the center of the petri dish and not close to the sidewall. Do not use too much adhesive. Let the adhesive dry. Remove excess adhesives using a sterilized scalpel or blade.
3. Clean the surface of the 3D printed mold and petri dish with ethanol and KimWipes.

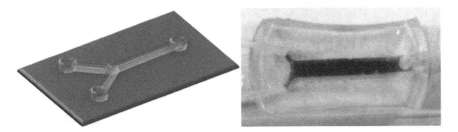

Fig. 11.18 Left: a 3D printed mold; right: an OOC made with replica molding

4. Instead of using PDMS gel and curing agent, we will use agarose gel. As we have learned in Chap. 6, agarose is a polysaccharide and much more cell-friendly than the strongly hydrophobic PDMS. Pour agarose solution into the petri dish, up to 1 cm above the petri dish surface. This depth (1 cm) is the thickness of the entire OOC. Wait for 5–10 min for the solution to solidify.
5. Using a sterilized scalpel or blade, remove the agarose replica from the mold on a petri dish. It will create a trench-shaped microfluidic chip shown in Figs. 11.3 and 11.4. Lift and detach the chip very carefully. If any deformation is observed, return to step 3 and repeat the process.
6. Using a biopsy punch, punch out two inlets and one outlet, as shown in Figs. 11.3 and 11.4.
7. Fabricate the bottom side chip using the same protocol.
8. Sandwich a porous membrane in between these two chips. Glue two chips together.

Question 11.3 You have determined to use a 3D printer to fabricate an OOC. Choose the best material for these 3D printed substrates and briefly justify your choice.

Objective 2. OOC Fabrication by Paper Microfluidics

In LOC, paper (= cellulose fiber) as a substrate and a *wax printer* as the printing method have become extremely popular as they are easy, low-cost, and fast (Chung et al., 2019; Ulep et al., 2020). Such paper-based LOC is called a *paper microfluidic chip*. A wax printer uses solid wax as its ink. Wax is temperately melted and transferred to the paper in a desired image or text, and it quickly returns to the solid state. As wax is strongly hydrophobic, it does not smear with water and provides superior printing quality than conventional inkjet or laser printing. For LOC applications, the inside channel is not printed with wax, where the liquid flows. The outside is filled with wax to create a hydrophobic barrier. Once the liquid is introduced inside the channel area, it spontaneously flows through the channel via *capillary action* (also known as *wicking*) (Fig. 11.19) (Chung et al., 2021).

Although paper microfluidic chips have rarely been used as OOCs, we can use them for OOC applications. Cellulose fibers can serve as good approximates of natural ECM. They are polysaccharides resembling GAGs and have fiber dimensions similar to collagen fibers (a few to a few tens of micrometer). Refer to Chap.

Fig 11.19 A paper
microfluidic chip

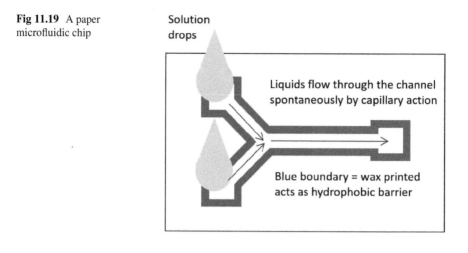

Solution
drops

Liquids flow through the channel
spontaneously by capillary action

Blue boundary = wax printed
acts as hydrophobic barrier

6 for details. However, it is preferable to modify the paper surface with ECM proteins, like collagens or laminins.

9. Design your paper microfluidic channel using SolidWorks or PowerPoint. One channel can be fabricated to mimic a single-channel structure or a bundle of channels can be manufactured. The example shown in Fig. 11.10 represents a liver lobule-like structure, with six sinusoid-like channels aligned circularly (Kaarj et al., 2020b). The top layer has six inlets (0.5 mm × 1.0 mm each; you can alter the dimensions), mimicking portal veins in the liver. Black-colored area is wax printed to serve as a hydrophobic barrier. Liquids will pass only through the white, non-wax-printed areas. The bottom layer has a six-channel structure (0.8 mm × 5 mm each; you can alter the dimensions), where the hepatocytes are anchored and proliferated. They flow through the central hole (1-mm diameter; you can change the size), mimicking a central vein in the liver.
10. Print the channel design on cellulose chromatography paper using a wax printer.
11. Cut the printed paper in the desired shape (circular shape for the liver-on-a-chip shown in Fig. 11.10).
12. Place the paper chip on a hot plate and set the temperature at 120°C. It will make the wax to be melted again and smear into the paper pores. This wax reflow creates a more robust hydrophobic barrier and prevents water leakage.
13. Place the paper chip inside the petri dish. Sterilize the paper chip and the petri dish under UV light inside a biosafety cabinet.
14. Add rat tail collagen type I to the paper chip. Follow Objective 3 of Laboratory Task 1 in Chap. 6. Collagens will be coated only to the areas where wax is not printed. Rinse with DPBS twice.

Objective 3. Cell Seeding on OOC
The next step is the cell seeing on OOC. You can use either primary rat hepatocyte (PRH) or HepG2 (immortalized human hepatocyte), both of which grow quite well in a laboratory environment. We will use HepG2.

Fig. 11.20 The liver-on-a-chip system

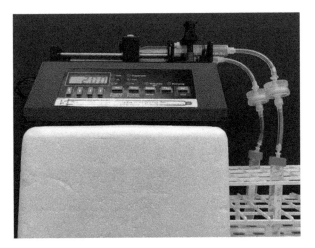

15. Centrifuge the cells at 1,500 RPM for 5 mins and resuspend the pellets at a final concentration of 10^6 cells/mL with appropriate medium, for example, Eagle's Minimum Essential Medium, supplemented with 10% v/v fetal bovine serum, 0.2% v/v of 250 µg/mL amphotericin B (antifungal), and 0.1% v/v of 50 mg/mL gentamycin sulfate (antibacterial).

16. Culture on a T-75 tissue culture flask at 37 °C in 5% CO_2 until 90% confluency. Passage cells as needed.

17. Place the bottom-layer paper chip into a petri dish. Add 3 mL of media. Seed 10^4 HepG2 cells on it. Incubate overnight at 37 °C in 5% CO_2.

18. After incubation, remove media and wash the paper chip twice with DPBS. HepG2 cells are anchored only to the inlets and six channels as all other areas are coated with wax.

19. Place the bottom- and top-layer paper chips within a syringe filter, as shown in Fig. 11.20.

11.15   Laboratory Task 2: Drug Testing with OOC

Objective 1. Drug Testing
The following drugs are tested. You can change the drugs of your choice.

– Phenacetin is a pain reliever and fever reducer. US FDA and many other countries currently ban it due to its toxicological effect. A recommended minimum daily dose is 1,200 mg (300 mg each four times a day). Since the average number of hepatocytes in humans is 1.68×10^{11} and there are 10^4 cells in our model, we can scale it down to 70 ng. Prepare 14.4 ng/mL of phenacetin and load 5 mL (= 70 ng) into the 10-mL syringe.

– Bupropion is an antidepressant. A recommended minimum daily dose is 75 mg. Using the same method, prepare 1 ng/mL bupropion and load 5 mL (= 5 ng) into the 10-mL syringe.

- Dextromethorphan is an antidepressant. A recommended minimum daily dose is 120 mg (20 mg each six times a day). Using the same method, prepare 1.4 ng/mL bupropion and load 5 mL (= 7 ng) into the 10-mL syringe.
- Use PBS (phosphate buffered saline) as a control.

20. Connect tubes as shown in Fig. 11.20. One is connected to a syringe pump (with a 10-mL syringe), where drug solutions are loaded, and the other to a waste container (15-mL tube).
21. Set the syringe pump at 2 μL/min (333 nL/min for each channel of the liver-on-a-chip model). The drug in the syringe is pumped and flew through the cells in the liver-on-a-chip model (within a syringe filter cartridge) for 40 min.
22. Disconnect the top tubing and collect the supernatant solution in a 15-mL tube. Disassemble the filter cartridge and take the bottom layer paper chip out.

Objective 2. Cell Counting

If a drug killed HepG2 cells, they would be detached from the surface. The number of HepG2 cells still bound on the paper chip is counted using DAPI staining and fluorescence microscopy.

23. Add 4% paraformaldehyde to fix the cells. Wash twice with 1 mL DPBS (Dulbecco's PBS).
24. Add 2 mL of DPBS followed by two droplets of NucBlue (DAPI) to stain nuclei. Allow sitting for 15 minutes under dark conditions (cover the petri dish with aluminum foil).
25. Image and count the number of cells within the paper chip with a fluorescence microscope with UV excitation.

Objective 3. Protein Production

Hepatocytes produce albumin. If a drug is toxic to the liver, such albumin production will be compromised. In this objective, a Bradford assay is used to quantify the total amount of proteins in the supernatant.

26. Warm the Bradford reagent to room temperature.
27. Mix the 50 μL supernatant with a 50 μL Bradford reagent. Incubate for 10 minutes.
28. Dilute the solution by tenfold by adding 900 μL DI water.
29. Transfer to a cuvette and measure the absorbance at 595 nm.
30. Separately prepare a series of bovine serum albumin (BSA) solutions with varying concentrations. Repeat the Bradford assay to construct a standard curve. Compare the data with the standard curve to calculate the total protein concentration.

Objective 4. Urea Production

Urea production is another popularly used measure in evaluating liver toxicity. In this object, bromocresol purple is used to quantify the total amount of urea in the supernatant.

31. Warm the bromocresol purple reagent to room temperature.
32. Mix the 20 μL supernatant with an 80 μL bromocresol purple reagent. Incubate for 20 min.
33. Dilute the solution by tenfold by adding 900 μL DI water.
34. Transfer to a cuvette and measure the absorbance at 588 nm.

35. Separately prepare a series of urea solutions with varying concentrations. Repeat the bromocresol purple assay to construct a standard curve. Compare the data with the standard curve to calculate the urea concentration.

Figure 11.21 shows the results of drug toxicity tests from the liver-on-a-chip. Gray bars represent the data under the flow condition, and black bars under the static condition (no flow). For both flow and static conditions, the cell counts are substantially lower with drugs than with control (PBS), indicating that the drugs are killing HepG2 cells. Similarly, both total protein production and urea production are compromised with drugs, more severely under the flow condition and phenacetin (currently banned).

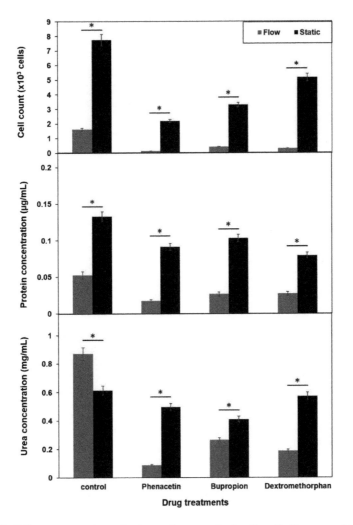

Fig. 11.21 Cell counts, protein production, and urea production from the liver-on-a-chip model. (Kaarj et al., 2020b. Reprinted with permission. (C) 2020 Springer)

Review Questions
1. Compare 2D and 3D cell cultures.
2. Compare 3D cell culture and organ-on-a-chip.
3. Compare photolithography and soft lithography for OOC fabrication.
4. Graphically illustrate how kidney-on-a-chip mimics the kidney.
5. Graphically illustrate how liver-on-a-chip mimics the liver.
6. Graphically illustrate how lung-on-a-chip mimics the lung.
7. Graphically illustrate how angiogenesis-on-a-chip mimics the angiogenesis.
8. Graphically illustrate how blood–brain-barrier-on-a-chip mimics the blood–brain barrier.
9. What is human-on-a-chip? Why is it necessary?
10. Why do you need gels in organ-on-a-chip?
11. What is ADME analysis?
12. How is pharmacokinetics different from pharmacodynamics?
13. What are the advantages of studying stem cell differentiation on organ-on-a-chip?
14. What are the advantages of studying metastasis on organ-on-a-chip?
15. What are the advantages of testing drug efficacy on organ-on-a-chip?
16. What are the advantages of evaluating drug/chemical/cosmetics toxicity on organ-on-a-chip?
17. How can you provide the flow (shear) to organ-on-a-chip? How can you generate programmed flow?
18. How can you provide compression or tension to organ-on-a-chip? How can you generate programmed compression or tension?
19. How can you create a chemical gradient on organ-on-a-chip?
20. Why is OOC better than microtiter plate in drug testing (efficacy and toxicity)?

References

Akarapipad, P., Kaarj, K., Liang, Y., & Yoon, J.-Y. (2021). Environmental toxicology assays using organ-on-chip. *Annual Review of Analytical Chemistry, 14*, 155–183. https://doi.org/10.1146/annurev-anchem-091620-091335

Cho, S., Islas-Robles, A., Nicolini, A. M., Monks, T. J., & Yoon, J.-Y. (2016). In situ, dual-mode monitoring of organ-on-a-chip with smartphone-based fluorescence microscope. *Biosensors and Bioelectronics, 86*, 697–705. https://doi.org/10.1016/j.bios.2016.07.015

Cho, S., & Yoon, J.-Y. (2017). Organ-on-a-chip for assessing environmental toxicants. *Current Opinion in Biotechnology, 45*, 34–42. https://doi.org/10.1016/j.copbio.2016.11.019

Chung, S., Breshears, L. E., Perea, S., Morrison, C. M., Betancourt, W. Q., Reynolds, K. A., & Yoon, J.-Y. (2019). Smartphone-based paper microfluidic particulometry of norovirus from environmental water samples at single copy level. *ACS Omega, 4*, 11180–11188. https://doi.org/10.1021/acsomega.9b00772

Chung, S., Breshears, L. E., Gonzales, A., Jennings, C. M., Morrison, C. M., Betancourt, W. Q., Reynolds, K. A., & Yoon, J.-Y. (2021). Norovirus detection in water samples at the level of single virus copies per microliter using a smartphone-based fluorescence microscope. *Nature Protocols, 16*, 1452–475. https://doi.org/10.1038/s41596-020-00460-7

Huh, D., Hamilton, G. A., & Ingber, D. A. (2011). From 3D cell culture to organs-on-chips. *Trends in Cell Biology, 21*, 745–754. https://doi.org/10.1016/j.tcb.2011.09.005

Huh, D., Kim, H. J., Fraser, J. P., Shea, D. E., Khan, M., Bahinski, A., Hamilton, G. A., & Ingber, D. E. (2013). Microfabrication of human organs-on-chips. *Nature Protocols, 8*, 2135–2157. https://doi.org/10.1038/nprot.2013.137

Kaarj, K., & Yoon, J.-Y. (2019). Methods of delivering mechanical stimuli to organ-on-a-chip. *Micromachines, 10*, 700. https://doi.org/10.3390/mi10100700

Kaarj, K., Madias, M., Akarapipad, P., Cho, S., & Yoon, J.-Y. (2020a). Paper-based in vitro tissue chip for delivering programmed mechanical stimuli of local compression and shear flow. *Journal of Biological Engineering, 14*, 20. https://doi.org/10.1186/s13036-020-00242-5

Kaarj, K., Ngo, J., Loera, C., Akarapipad, P., Cho, S., & Yoon, J.-Y. (2020b). Simple paper-based liver cell model for drug screening. *BioChip Journal, 14*, 218–229. https://doi.org/10.1007/s13206-020-4211-6

Liang, Y., & Yoon, J.-Y. (2021). In situ sensors for blood-brain barrier (BBB) on a chip. *Sensors and Actuators Reports, 3*, 100031. https://doi.org/10.1016/j.snr.2021.100031

Openstax. (2013). *Anatomy and physiology.* http://cnx.org/content/col11496/1.6/

Ulep, T.-H., Zenhausern, R., Gonzales, A., Knoff, D. S., Lengerke Diaz, P. A., Castro, J. E., & Yoon, J.-Y. (2020). Smartphone based onchip fluorescence imaging and capillary flow velocity measurement for detecting ROR1+ cancer cells from buffy coat blood samples on dual-layer paper microfluidic chip. *Biosensors and Bioelectronics, 153*, 112042. https://doi.org/10.1016/j.bios.2020.112042

Yoon, J.-Y. (2016). *Introduction to biosensors: from electric circuits to immunosensors* (2nd ed.). New York, Chapter 14. https://doi.org/10.1007/978-3-319-27413-3_14

Chapter 12
Tissue-Engineered Skin Transplant

There are numerous examples of tissue-engineered (TE) transplants. TE skin transplant is perhaps the oldest and the most straightforward example of them. In this chapter, we will learn

– The basic anatomy and physiology of human skin
– When TE skin transplant is needed
– Current methods of TE skin transplants

Inquiry 1. Skin is one of the most prolific tissues in the human body. Can you explain why?

Inquiry 2. When do you think you need a TE skin transplant?

12.1 When Do You Need Skin Transplants?

Skin is the first defense line, protecting the human from various pathogens (viruses, bacteria, fungi, etc.). When you have a cut or a small wound in the skin, you have noticed that it healed itself relatively fast and left no scar in most cases. Skin is highly prolific and self-renewable, and early tissue engineers have paid attention to the tissue-engineered skin transplants as one of the early tissue engineering examples.

Damages made to a small cut or a small wound are typically restricted to the epidermis. Hence, they are healed relatively quickly, leaving no scar. However, severe burns do not heal very well, as the damages are done deep into the dermis layer (Fig. 12.1). Scars are expected in severe burns. In this case, a transplant may become necessary.

Diabetic patients also suffer from skin ulcers that do not heal well, called *diabetic ulcers* (Fig. 12.2). Diabetic ulcers are prominent in the foot as it is the farthest part of the body from the heart. This problem is called a *diabetic foot*. Again, a skin transplant may be a solution.

© Springer Nature Switzerland AG 2022
J.-Y. Yoon, *Tissue Engineering*, https://doi.org/10.1007/978-3-030-83696-2_12

Fig. 12.1 Third-degree
burn on foot. The picture
was taken by Craig0927 in
November 2009 and placed
in the public domain.
(Accessed March 2021
from https://commons.
wikimedia.org/wiki/
File:8-day-old-3rd-degree-
burn.jpg)

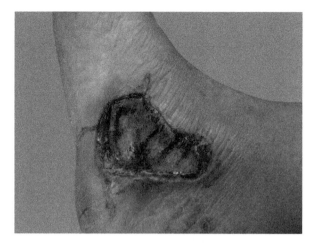

Fig. 12.2 Diabetic foot
ulcer. (Image by Mark
A. Dreyer, DPM,
FACFAS – https://www.
ncbi.nlm.nih.gov/books/
NBK537328/figure/
article-34555.image.
f1/?report=objectonly, CC
BY 4.0, https://commons.
wikimedia.org/w/index.
php?curid=97354949)

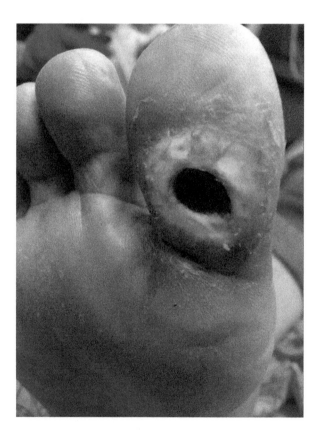

Skin transplant can also be used for plastic surgery to replace the aged skin with a fresh one.

12.2 Basic Anatomy of Skin

We have already covered the basic anatomy of the skin in Chap. 1. Let us revisit one more time here. The skin has two layers: (1) *epidermis*, the outermost and top layer, and (2) *dermis*, the immediate inner layer (just underneath the epidermis). *Keratinocytes* and *fibroblasts* make up most of the epidermis and dermis, and they proliferate well both in vivo and in vitro. See Fig. 12.3 (identical to Fig. 1.4) for the structure of the skin.

As shown in Fig. 12.3, keratinocytes move slowly from the bottom (inner layer) to the top (outer layer), becoming *squames* (*keratin scales*). The keratinocytes in the innermost layer are cuboidal-shaped, sitting on a basement membrane, while they are flattened as they move up to the outer layer. The doubling time of human keratinocytes is approximately 50–70 days, which may not be fast enough to be cultured in vitro.

Like other cells, keratinocytes have three types of cytoskeletons: actin filaments, microtubules, and intermediate filaments. *Keratin fiber* is one type of intermediate filament found in many mammalian cells and playing an essential function in

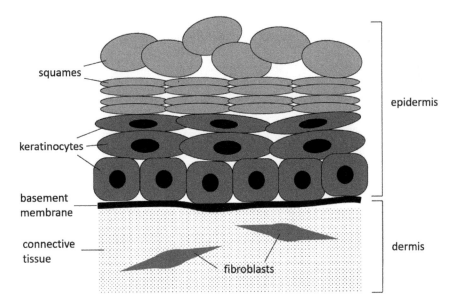

Fig. 12.3 (Identical to Fig. 1.4) Human skin. The outermost layer is the epidermis (consisting of squames and keratinocytes), followed by the dermis (consisting of fibroblasts and ECM).

keratinocytes, as its name suggests. Keratin fiber is also a significant component of human hair and nails.

The boundary of the epidermis and dermis forms the *dermoepidermal junction* (*DEJ*). This boundary should be connected to prevent the epidermis' separation from the dermis, preventing cells from passing through while allowing small molecules (oxygen, nutrients, etc.) to pass through. The basement membrane between the epidermis and dermis is the crucial component of this DEJ.

12.3   How Can We Culture Keratinocytes In Vitro?

While skin is a prolific tissue in the human body, keratinocytes (the major cell component in the epidermis) do not proliferate rapidly in vitro. The first in vitro culture of keratinocytes was demonstrated using fibroblasts as a feeder layer, like the stem cell culture (refer to Chap. 5). This demonstration is similar to the in vivo condition, where the fibroblasts (along with the connective tissue made by the fibroblasts) form the dermis layer just underneath the epidermis layer. Since then, numerous protocols have been developed to culture keratinocytes in vitro. Recently, a newer method of in vitro keratinocyte culture has been developed that does not require the feeder (i.e., fibroblast) layer.

Keratinocytes can be harvested from a donor human or the patient himself/herself. However, depending on the age, the number of doublings can be limited (Hayflick limit). The resulting keratinocyte culture forms either a monolayer or a multilayered sheet, depending on the culture condition. Since the cultured keratinocyte sheets are quite fragile, they require a fibrous membrane structure (i.e., simulating the basement membrane) made from synthetic polymers.

12.4   Langerhans Cells and Immune Response of Skin

In the epidermis, there are *Langerhans cells,* although their number is much smaller than that of the keratinocytes. Langerhans cells are the skin-tissue-specific macrophages. Together with the neutrophils (one type of white blood cells) in the blood vessel, macrophages *phagocytose* any "foreign" substances such as bacteria, viruses, cancer cells, cellular debris, etc., in the tissue. Macrophages are initially monocytes in the blood vessel, thus one type of white blood cells, which become macrophages when they escape from the blood vessel. As the skin is the first defense line, they must reject the foreign substances physically and immunogenically. Langerhans cells play the latter role.

However, Langerhans cells do not proliferate well in vitro. Langerhans cells in the tissue-engineered (TE) transplant can cause an unnecessary immune response. Therefore, they are not deliberately added to the in vitro culture of TE skin transplants.

12.5   Scaffold for Tissue-Engineered Skin Transplant

While it is possible to proliferate keratinocytes in vitro, it is preferred to have a fibroblast feeder layer to construct a healthy keratinocyte sheet. Another important aspect is the recapitulation of the basement membrane. It provides the anchorage surface to the keratinocytes and the structural integrity (tensile strength, elasticity, and firm bonding between epidermis and dermis) of the skin transplant. Hence, the basement membrane mimic is necessary as a TE scaffold.

As briefly introduced in Chap. 1, the decellularized matrix (Sect. 1.2) has also been utilized toward the TE skin transplant. This decellularized skin matrix contained the basement membrane and other extracellular matrix components (most notably the connective tissue making up the inner dermis layer). However, as explained in Chap. 1, there are many problems using the decellularized matrix as tissue-engineered scaffold. These problems include (1) lack of control on its structure and morphology, (2) availability of cadavers (corpses), and (3) presence of bacteria.

The following is a list of various TE skin scaffolds:

1. *Decellularized matrix*: *AlloDerm* is probably the earliest example of the TE skin scaffold. It is essentially the decellularized skin matrix from a human cadaver or an animal. Decellularization of skin matrix is typically conducted by freeze-drying, preserving the basement membrane that separates the epidermis and the dermis. AlloDerm has popularly been used for plastic surgeries (subcutaneous filler, breast reconstruction, nose plastic surgery, etc.) and dental surgeries. However, being the decellularized matrix, it has many limitations, as mentioned above (Fig. 12.4).

2. *Polymer membrane*: *Dermagraft* is a polymer membrane (made from, e.g., bio-degradable polymer PGA) seeded with neonatal (i.e., from a newborn human) fibroblasts from a donor. This polymer scaffold contains growth factors, ECM proteins (e.g., collagen, fibronectin, vitronectin, laminin, etc.), and glycosamino-glycans (GAGs). It works well with severe burns where the dermis was damaged. However, it reconstructs only the dermis but not the epidermis (the latter occurs naturally by the human body), the clinical outcome is somewhat limited (Fig. 12.5).

3. *Collagen + GAG gel*: A better approach is the extracellular matrix's recapitulation with collagen fibers and glycosaminoglycans (GAGs). Therefore, gels made from collagen fibers or GAGs have later been utilized as the scaffold for the TE skin transplant. As they contain >90% water, these gels are called hydrogels. Among GAGs, chondroitin sulfate (CS) and hyaluronic acid (HA) have been used to create a hydrogel toward the TE skin transplant. Chitosan-based hydro-gel has also been used due to its similarity to the GAGs. Electrospinning (discussed in Chap. 9) is another possibility of creating a hydrogel as the TE skin scaffold. As the skin has two layers – epidermis and dermis – it is preferred to have a two-layered TE skin scaffold, proliferated with keratinocytes and fibro-blasts, respectively. *Integra Dermal Regeneration Template* is made from the

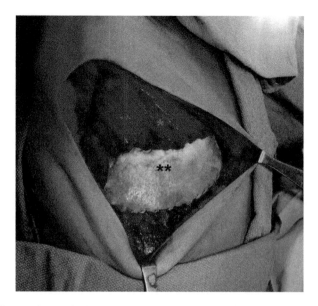

Fig. 12.4 AlloDerm is used for breast prosthesis. (Becker et al., 2009. Reprinted with permission, (C) 2009 Wolters Kluwer Health)

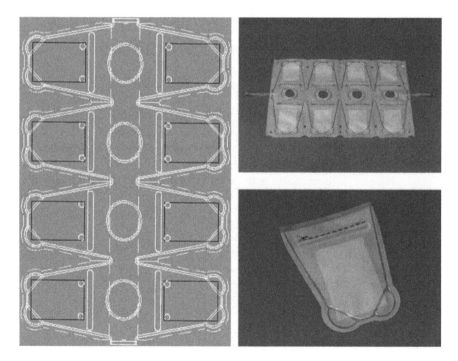

Fig. 12.5 Dermagraft are placed in a bioreactor to proliferate cells on it. (Mansbridge, 2006. Reprinted with permission, (C) 2006 John Wiley and Sons)

collagens from cattle and the chondroitin sulfate (one type of GAGs) from sharks, covered with a silicone layer. It has been used to treat wounds. Fibroblasts and vascular endothelial cells from the nearby tissue can migrate toward this scaffold, creating the dermis layer (with fibroblasts) and the blood vessels (with vascular endothelial cells). After three weeks, a separate thin sheet is added to promote epidermis formation (with keratinocytes from the nearby tissue) (Fig. 12.6).

4. *Collagen + GAG gel with pre-cultured cells*: Rather than depending on the cells from the nearby tissues, the collagen-GAG membrane can be pre-seeded with keratinocytes or fibroblasts and proliferated in vitro. Both keratinocytes and fibroblasts can be harvested from the same patient (autologous cells). *Permaderm*

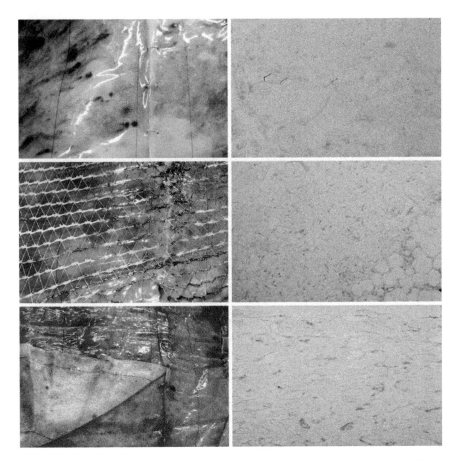

Fig. 12.6 Integra Dermal Regeneration Template is added to a scarred abdominal skin. Clinical appearances are in the left column and histological images in the right column. First row = 0 day, second row = 21 days, and third row = 4 weeks. (Moiemen et al., 2006. Reprinted with permission, (C) 2006 Wolters Kluwer Health)

Fig. 12.7 (Identical to Fig. 1.5) Bilayered tissue-engineered skin (MyDerm). (Ude et al., 2018. (C) Open access article distributed under the terms of the Creative Commons Attribution 4.0 International License)

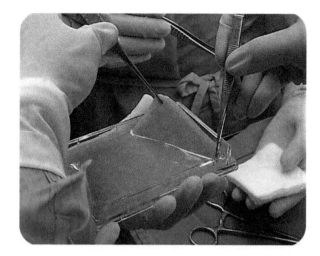

is one example of collagen + GAG gel with pre-cultured cells. *MyDerm*, briefly explained in Chap. 1, is another example of a bilayered skin scaffold with pre-cultured cells (Fig. 12.7).

12.6   Use of Stem Cells for TE Skin Transplant

While fibroblasts proliferate relatively well under in vitro culture, keratinocytes proliferate slowly due to their quite long doubling time (50–70 days). Therefore, it will be beneficial to use keratinocyte stem cells than somatic keratinocytes. This process is how a wound is healed under in vivo conditions.

Keratinocyte stem cells (*KSCs*) can be found around the basement membrane, that is, the dermoepidermal junction (DEJ). This junction is the boundary between the epidermis and dermis. However, it is somewhat challenging to locate the KSCs from the patient.

Induced pluripotent stem cells (iPSCs) can also be used. Somatic keratinocytes or somatic fibroblasts can be converted from iPSCs by adding transcription factors, such as Oct3/4, Sox2, c-Myc, and Klf4, as we have learned in Chap. 5. Mesenchymal stem cells (MSCs) can also be used. Both iPSC and MSC can be differentiated into keratinocytes and fibroblasts, and if needed, vascular endothelial cells (to construct blood vessels in the skin transplant). With proper morphogenetic factors (growth factors, differentiation factors, physical stimuli, adequate environment, etc.), these stem cells can be used to reconstruct the skin tissue around the TE skin scaffold. However, "correct" tissue development has always remained a significant challenge.

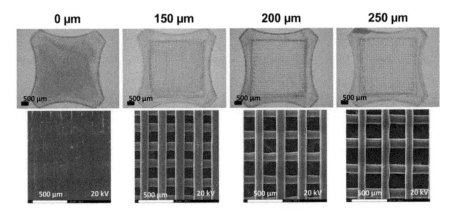

Fig. 12.8 3D printed PLA scaffolds. (Gremare et al., 2018. Reprinted with permission, (C) 2018 John Wiley and Sons)

12.7 Laboratory Task 1: Construction of Skin TE Scaffold

In this task, we will construct a mesh-shaped skin TE scaffold made from polylactic acid (PLA, biodegradable polymer) using a 3D printer. This procedure is adapted from Gremare, et al., 2018. While it was originally intended for bone tissue engineering, it is adapted for skin tissue engineering in this laboratory task.

1. Using appropriate software (e.g., SolidWorks), design the PLA mesh as shown in Fig. 12.8. Vary the mesh dimensions from 150 μm to 250 μm, or any number the 3D printer is capable of printing.
2. Using the 3D printer, print the mesh using the PLA material.

It is also possible to secure the polymer membrane from the commercial vendors described in this chapter, for example, AlloDerm, Dermagraft, Integra, Permaderm, MyDerm, etc.

12.8 Laboratory Task 2: Seeding and Proliferating Keratinocytes on the TE Skin Scaffold

In this task, we will seed and culture keratinocytes or fibroblasts on the PLA mesh scaffolds. When using keratinocytes, it is essential to use the keratinocytes with a short doubling time.

3. Place the PLA mesh within the 12-well or 24-well plate.
4. Sterilize each well under 70% ethanol for 10 min and rinse three times with PBS.
5. Soak each well in the growth media without antibiotics for 12 h to allow proteins to attach to the PLA mesh. Refer to the laboratory tasks of Chap. 3 for details.

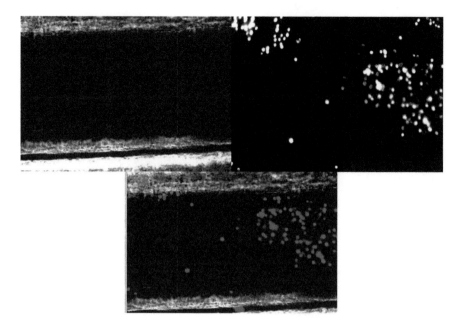

Fig. 12.9 Cells cultured on the PLA mesh. Top left shows the bright field microscopic image of the PLA mesh with cultured fibroblasts. Top right is the fluorescence microscopic image of the same with DAPI staining. Bottom shows the overlayed image with blue pseudo-color applied to the DAPI-stained image.

6. Seed mouse fibroblasts or rat keratinocytes from newborn rats (or other appropriate keratinocytes whose doubling time is on the scale of a few days) on each well with the cell density of 5×10^5 cells/cm^2.
7. Incubate the cells for two days to allow the complete adhesion of cells and allow them to reach confluency (submerged in media within the well plate).
8. Remove the scaffold with the cultured cells from the well plate.
9. Place it on a glass slide. Image the cell adhesion on the scaffold under a light microscope. Stain with DAPI, anti-vinculin, phalloidin, etc., as necessary (Fig. 12.9).

12.9   Laboratory Task 3: Force: Deflection Curve of the TE Skin Transplant

Finally, the TE skin graft (the scaffold with the cultured fibroblasts or keratinocytes) is bent by hand (*strain*), and a simple spring scale measures the force (*stress*) needed to make such elongation. A *spring scale* is typically used to measure a mass's weight, while we attach this spring scale to the TE skin graft to measure the force (stress) applied to the TE skin transplant (Fig. 12.10).

Fig. 12.10 A spring scale

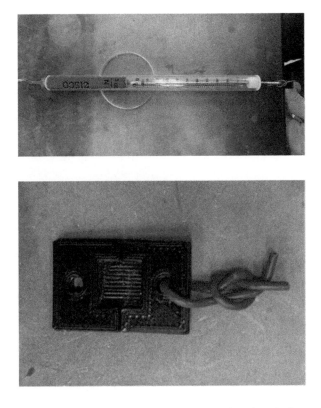

Fig. 12.11 A wire is attached to the TE skin transplant. You can notice the PLA mesh at the center, as well as the traces of cells on it.

10. Carefully remove the transplant from the media. Dry the transplant by placing KimWipes and allowing to air-dry for 1 minute. Wipe away remaining media on the edges.
11. As a control, prepare the bare PLA mesh, that is, without cells.
12. Attach wire to the transplant (and the bare PLA mesh) (Fig. 12.11).
13. Connect it to a spring scale. Hold the other edge on the table (Fig. 12.12).
14. One person pulls the scale, the other person holds the transplant, and the third takes images. The transplant is bent. Read the scale to obtain the applied force. The deflection distance is measured from the images (Fig. 12.13).
15. Plot the force–deflection curve.

Table 12.1 shows the sample data set, and Fig. 12.14 shows the resulting force–deflection curve. Actual data will vary significantly by the material (PLA), the 3D printer, the cell type, and the culture conditions. This curve in no way represents the stress–strain curve, although the concept is pretty similar. Stress is the force per unit area and hence has a unit of pressure (Pa), and strain is the percent deformation divided by the original dimension (e.g., length). Since the transplant is bent, it represents bending stress and bending strain.

In the classic stress–strain curve, the initial slope represents *modulus*, a ratio of stress over strain. If the material is elongated, it is *elastic modulus* or *Young's*

Fig. 12.12 A TE skin
transplant is hooked to the
spring scale

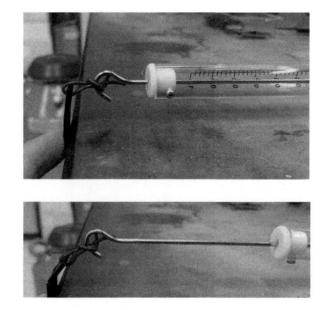

Fig. 12.13 Deflection
distance is measured
(yellow arrow)

Table 12.1 Strain and stress data of TE skin scaffold without and with cells cultured

PLA mesh (no cells)		PLA mesh + fibroblasts	
Deflection (mm)	Force (N)	Deflection (mm)	Force (N)
0	0	0	0
0.71	0.1	0.06	0.1
1.14	0.2	0.43	0.2
1.58	0.3	0.76	0.3
2.35	0.4	0.87	0.4
		1.66	0.5
		1.97	0.6
		2.21	0.7

modulus. It is a measure of elasticity, showing the material's ability to deform (elongated) under a given stress. In our case, it is *bending modulus*, representing the material's ability to bend under a give stress. You can see that the slope (corresponding to the bending modulus) of the TE skin transplant, that is, with cells, is steeper than that without cells, indicating superior material property. In other words, the transplant resists the external force better than the one without cells.

Typically, there is a maximum point of such curve, where the maximum stress is called *yield strength*. Until the maximum strength point, the material will be deformed elastically, that is, the deformation returns to the original status without any permanent deformation. This deformation is called *elastic deformation*. After passing the maximum strength point, the deformation cannot return to the original status – some of the deformations become permanent. This deformation is called *plastic deformation*. Applying additional stress or strain beyond the yield strength

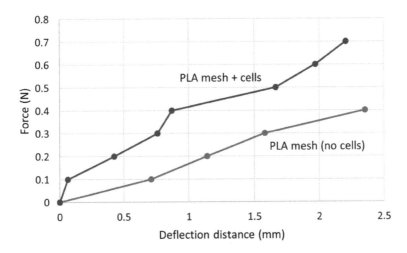

Fig. 12.14 Force–deflection curves of TE skin transplant, without and with cells cultured

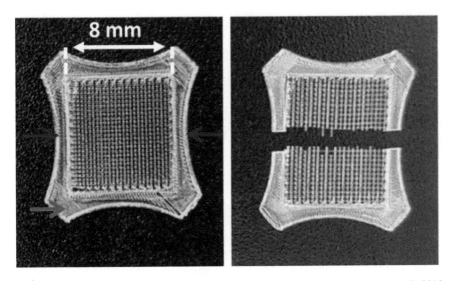

Fig. 12.15 Fractured TE skin scaffold (Gremare et al., 2018. Reprinted with permission, (C) 2018 John Wiley and Sons)

causes plastic deformation until the material eventually fractures. The stress where the fracture occurs is called *fracture strength* (Fig. 12.15). You can integrate the stress–strain curve up to this fracture strength point, representing the material's toughness. Therefore, various materials properties can be obtained from a stress–strain curve, as summarized in Fig. 12.16.

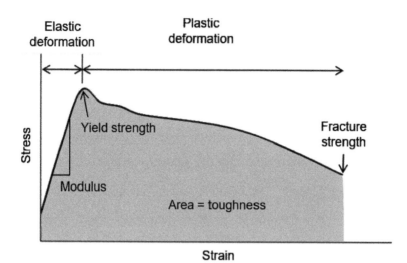

Fig. 12.16 Stress–strain curve, with various material properties, including modulus, yield strength, fracture strength, toughness, and elastic versus plastic deformation

Review Questions

1. What are diabetic ulcers and diabetic foot?
2. Explain the two layers of the skin and the primary cells in each layer.
3. What is dermoepidermal junction (DEJ)?
4. Compare the differences between culturing keratinocytes and fibroblasts.
5. What are Langerhans cells and what is their function in the skin?
6. Compare the following four TE skin scaffolds: (1) decellularized matrix, (2) polymer membrane, (3) hydrogels (collagens and/or GAGs), and (4) hydrogels with the pre-cultured cells.
7. Why do you need stem cells in skin tissue engineering?
8. Discuss the strengths and weaknesses of keratinocyte stem cells (KSCs), iPSCs, and mesenchymal stem cells (MSCs) in skin tissue engineering.
9. Define elastic modulus, yield strength, fracture strength, and toughness from the stress–strain curve. What are the physical meanings of them?

References

Becker, S., Saint-Cyr, M., Wong, C., Dauwe, P., Nagarkar, P., Thornton, J. F., & Peng, Y. (2009). AlloDerm versus DermaMatrix in immediate expander-based breast reconstruction: a preliminary comparison of complication profiles and material compliance. *Plastic and Reconstructive Surgery, 123*, 1–6. https://doi.org/10.1097/PRS.0b013e3181904bff

Gremare, A., Guduric, V., Bareille, R., Heroguez, V., Latour, S., L'heureux, N., Fricain, J. C., Catros, S., & le Nihouannen, D. (2018). Characterization of printed PLA scaffolds for bone tissue engineering. *Journal of Biomedical Materials Research Part A, 106A*, 887–894. https://doi.org/10.1002/jbm.a.36289

Mansbridge, J. (2006). Commercial considerations in tissue engineering. *Journal of Anatomy, 209*, 527–532. https://doi.org/10.1111/j.1469-7580.2006.00631.x

Moiemen, N. S., Vlachou, E., Staiano, J. J., Thawy, Y., & Frame, J. D. (2006). Reconstructive surgery with Integra Dermal Regeneration Template: Histologic study, clinical evaluation, and current practice. *Plastic and Reconstructive Surgery, 117*, 160S–174S. https://doi.org/10.1097/01.prs.0000222609.40461.68

Ude, C. C., Miskon, A., Idrus, R. B. H., & Bakar, M. B. A. (2018). Application of stem cells in tissue engineering for defense medicine. *Military Medical Research, 5*, 7. https://doi.org/10.1186/s40779-018-0154-9

Chapter 13
Vascularization of Tissue Transplants

So far, we have covered the proliferation of cells and the fabrication of scaffolds, and two examples of tissue engineering applications. To provide necessary nutrients, oxygens, and other chemicals, you must add blood vessels to the tissue-engineered devices. OOCs do not require such addition of blood vessels (= *vascularization*), as nutrients, oxygens, and chemicals are provided from an external pump. For tissue-engineered transplants, however, vascularization is necessary to prevent cell death. Blood vessels can be naturally formed from the nearby tissues upon transplantation, and it may not require deliberate vascularization. However, it is generally preferred to have an intentional vascularization protocol.

Inquiry 1. Have you heard of vascular endothelial cells?

Inquiry 2. Have you heard of angiogenesis?

13.1 Angiogenesis and Vascularization

Angiogenesis (angio = vessel and genesis = origin or formation) is a physiological process involving the growth of new blood vessels from preexisting vessels (Fig. 13.1). Angiogenesis is a normal process in tissue growth and development. It also plays a critical role in wound healing. Unfortunately, a high extent of angiogenesis is also a signal for a tumor to turn from a dormant (= benign) state to a malignant state (= cancer) as cancer cells are "crazy" and consume an abnormally high amount of nutrients and oxygen.

In normal tissue development, angiogenesis starts the formation of a blood vessel network. Such network formation is finished when the blood vessel network fills the entire tissue to provide necessary perfusion and tissue-specific functions. Such a process is called *vascularization*. *Neovascularization* refers to forming a new blood vessel network, while *revascularization* to reforming the blood vessel network previously destroyed.

© Springer Nature Switzerland AG 2022
J.-Y. Yoon, *Tissue Engineering*, https://doi.org/10.1007/978-3-030-83696-2_13

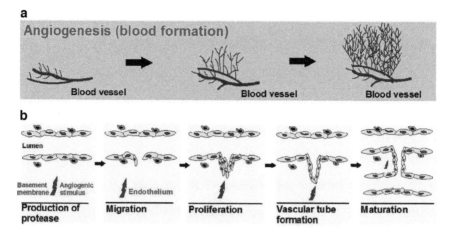

13.2   Anatomy and Physiology of Blood Vessel

A blood vessel is essentially a cylindrical tube made from cells and extracellular matrix (ECM) like other tissues. The inside of a hollow tube is called the *lumen*, where blood flows. The first cells in direct contact with blood are *vascular endothelial cells* (*VECs*). VECs sit on top of a thin ECM layer (of course through focal adhesion), called the *internal elastic lamina*, primarily made from *elastin*. Beyond internal elastic lamina, there are multilayers of *vascular smooth muscle cells* (*VSMCs*), whose thickness varies significantly by the type of blood vessels (some blood vessels do not have smooth muscle cells). These vascular smooth muscle cells are surrounded by connective tissue made by fibroblasts and ECM. Again, the thickness of this connective tissue varies significantly by the type of blood vessels (some blood vessels do not have this connective tissue layer).

Anatomically speaking, the vascular endothelial cells and the internal elastic lamina form *tunica intima* (tunica = fabric or cloth; intima = innermost) or simply *intima*. The vascular smooth muscle cell layer is called *tunica media* (media = middle) or simply *media*. The connective tissue layer is called *tunica adventitia* (adventitia = outermost) or simply *adventitia*. Figure 13.2 depicts the anatomy of a blood vessel.

When a new blood vessel develops, called a *nascent vessel*, it is merely a tube of vascular endothelial cells and is not an actual blood vessel. The addition of internal elastic lamina turns this nascent vessel into a functioning *capillary vessel* or simply *capillary*. Capillary vessels are also wrapped with a very thin and sparse layer of

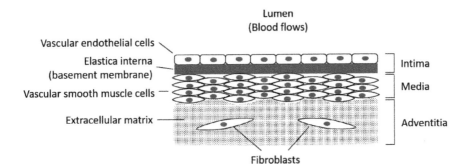

Fig. 13.2 Anatomy of a blood vessel

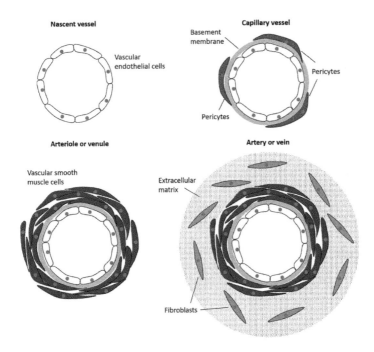

Fig. 13.3 Anatomy of the nascent vessel, capillary vessel, arteriole/venule, and artery/vein

pericytes. Capillaries can be further enlarged with vascular smooth muscle cells (VSMCs) and become small arteries or small veins. They are respectively called *arterioles* and *venules*. Further connective tissue will make the vessel into normal *arteries* and *veins*, which have all three layers (intima, media, and adventitia) and all three cell types (VECs, VSMCs, and fibroblasts) (Fig. 13.3).

13.3 Process of Angiogenesis

Angiogenesis is triggered by both chemical factors [most notably by *vascular endothelial growth factor (VEGF)*] and mechanical stimulations (shear force and compression). Chemical factors bind to the receptor on the vascular endothelial cells (VECs), for example, the *VEGF receptor (VEGFR)*. The receptors in the VECs can also recognize mechanical stimulations. Once such binding occurs, the basement membrane (e.g., internal elastic lamina) in the blood vessel is degraded, and the VECs start proliferating. Since the inner surface of a blood vessel is saturated with VECs and the basement membrane is already degraded, the newly divided VECs sprout outside the vessel, as shown in Fig. 13.4. This process is called *vessel sprouting*. This vessel continues to extend until it connects to the other vessel, toward forming a network. This nascent vessel is added with the internal elastic lamina, pericytes, vascular smooth muscle cells, etc. They will eventually develop into capillaries, arterioles, venules, etc., and subsequently stabilized.

Arteries must be connected to the veins toward a well-functioning vessel network. Physiologically, this process is coordinated by the ephrin on arteries and the Eph receptor on veins. On the arteries, *ephrin ligand* or simply *ephrin* is expressed on its surface, which can bind to the *Eph receptor* expressed on the surface of veins. For example, ephrin-B2 on arteries bind to EphB3 and EphB4 on veins (Fig. 13.5).

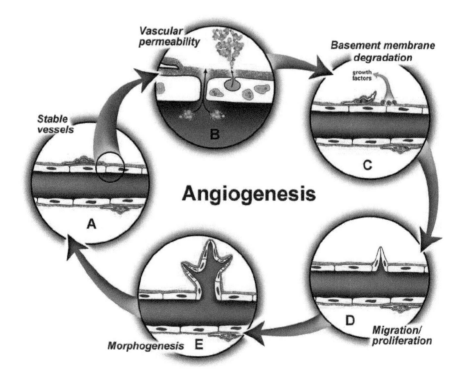

Fig. 13.4 Process of angiogenesis. (Bryan & D'Amore, 2007. Reprinted with permission, (C) 2007 Springer Nature)

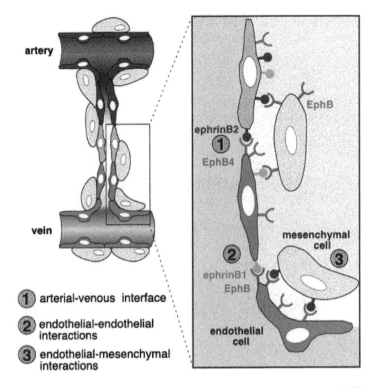

Fig. 13.5 Ephrin and Eph receptor on the arterial-venous boundary. (Adams & Klein, 2000. Reprinted with permission, (C) 2000 Elsevier)

Therefore, the presence of ephrin–Eph receptor binding indicates the boundary between arterial and venous domains. This ephrin–Eph receptor interaction prevents intermixing of arterial capillary and venous capillary, especially during the sprouting and extension of nascent vessels for angiogenesis. This process leads to the correct formation of a capillary network.

13.4   Angiogenesis-Stimulating Growth Factors and Angiogenesis Inhibitors

A healthy body controls angiogenesis through a series of on and off switches, and we can utilize those switches toward tissue engineering.

The "on" switches are angiogenesis-stimulating growth factors. We have already learned the most prominent example, *vascular endothelial growth factor* (*VEGF*), which increases the number of capillaries in a vasculature. Exercise leads to muscle contractions, increased blood flow, an increase in mRNA production for VEGF and VEGF receptors, and finally, VEGF and VEGF receptor production toward angiogenesis. *VEGF receptor* is the receptor found on, for example, vascular endothelial cells during the angiogenesis process. There is a co-receptor for the VEGF receptor, called *neuropilin-1*.

Angiopoietin-1 (Ang-1) similarly promotes angiogenesis. Its receptor is *tyrosine kinase with immunoglobulin-like and EGF-like domains 2, or Tie-2* for short.

Fibroblast growth factor (FGF) is a general growth factor for many connective tissues, and it can also stimulate the growth of vascular endothelial cells (VECs), vascular smooth muscle cells (VSMCs), and fibroblasts. FGF-2 is the most well-studied example of general tissue growth.

Platelet-derived growth factor (PDGF) is another growth factor primarily involved in blood coagulation and subsequent wound healing. PDGF triggers tissue growth and vascularization. During angiogenesis, PDGF specifically promotes VSMC proliferation. Similarly, *plasminogen activator* can function as an angiogenesis stimulator. When blood clots form, they need to be destroyed after some time, called *fibrinolysis* (fibrin = blood clots; lysis = destruction). There is *plasminogen* in the blood that can potentially destroy blood clots, although it is not an active form. Plasminogen activator signals the conversion of plasminogen to *plasmin*, which can destroy blood clots. Plasminogen activator can also signal the release of the growth factors involved in angiogenesis.

Transforming growth factor β (TGF-β) increases extracellular matrix (ECM) production and stimulates angiogenesis.

All these factors, except one, stimulate the growth of many different tissues, while VEGF is specific to blood vessels (and angiogenesis). Therefore, tissue engineers have paid their attention to VEGF to engineer vascularization of tissue-engineered devices.

The "off" switches are angiogenesis inhibitors. Before tissue engineering applications, scientists have paid their attention to these angiogenesis inhibitors to treat cancer. As cancer cells are "crazy," they consume abnormally high amounts of nutrients and oxygen with deregulated metabolism. As a result, cancer cells demand more blood vessels, and an increased level of angiogenesis is usually an indication of a malignant tumor, that is, cancer.

VEGF receptor (VEGFR) is commonly upregulated during angiogenesis. However, the addition of an ample number of VEGF receptors and/or its co-receptor *neuropilin-1* to blood or tissue ECM can attenuate angiogenesis. In this manner, they can function as angiogenesis inhibitors, called decoy receptors. (Note that sufficient expression of VEGFR on the vascular endothelial cells works toward stimulating angiogenesis, provided that there are sufficient VEGF.)

Angiopoietin-2 (Ang-2) is an antagonist to angiopoietin-1 (Ang-1), and thus destroys vascularization.

Angiostatin, endostatin, and *thrombospondin (TSP*, including TSP-1 and TSP-2) are other well-known angiogenesis inhibitors. Endostatin similarly functions as an angiogenesis inhibitor.

Plasminogen activator inhibitor-1 (PAI-1), also known as *serpin E1*, controls blood clot destruction (fibrinolysis). PAI-1 is used as an angiogenesis inhibitor. (Note that plasminogen activator is used as an angiogenesis stimulator.)

There are many other examples of angiogenesis inhibitors, including vasostatin, calreticulin, prolactin, platelet factors, interferons (IFNs), certain interleukins (ILs), secreted protein acidic and rich in cysteine (SPARC), etc. Many of them are discovered as potential drugs for cancer through inhibiting blood supplies to malignant tumors.

Antibody to VEGF (anti-VEGF) can also be used as an angiogenesis inhibitor, although they are not naturally found in the human body. They must be synthesized using *hybridoma* cells: the fused cells of normal B lymphocytes and cancer B lymphocytes. The most well-known example of anti-VEGF is *bevacizumab* (commercial name Avastin) and has successfully been used to treat various cancers.

Question 13.1 List all angiogenesis stimulators and inhibitors involved in normal blood coagulation and subsequent would healing process. Identify angiogenesis stimulators and inhibitors.

Question 13.2 When does VEGFR work as an angiogenesis stimulator? When does it work as an angiogenesis inhibitor?

13.5 HUVEC

Somatic vascular endothelial cells do not proliferate very well in vitro conditions. Because of this, *human umbilical vein endothelial cells (HUVECs)* have popularly been used for laboratory research and organ-on-a-chip applications. As its name suggests, HUVECs are isolated from an umbilical cord, typically discarded after birth. HUVECs proliferate quite well under in vitro conditions.

13.6 Formation of Vasculature Network

Angiogenesis-stimulating growth factors, especially VEGF, promote capillary growth, ultimately creating a network of capillary vessels or vasculature. We can also duplicate such vasculature formation in vitro. Figure 13.6 shows one such example. HUVECs are grown on Matrigel (solubilized basement membrane enriched with laminin). The addition of VEGF (right) promotes the creation of a vasculature network.

Bone morphogenetic protein 4 (BMP-4) belongs to the family of TGF-β and is involved in bone and cartilage development. It is also involved in stem cell differentiation. Despite its name, BMP-4 is also involved in the formation of a vasculature network. Figure 13.7 shows the results of human embryonic stem cell (hESC) culture on Matrigel, without or with 100 ng/mL BMP-4. BMP-4 treatment increases the outgrowth of network structures.

Fig. 13.6 HUVECs are grown on Matrigel, creating a vasculature structure. (Pill et al., 2015. (C) 2015 Pill et al. Open access article distributed under the terms of the Creative Commons Attribution 4.0 International License)

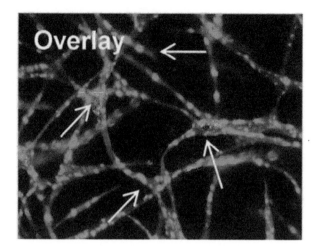

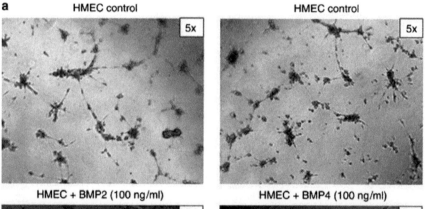

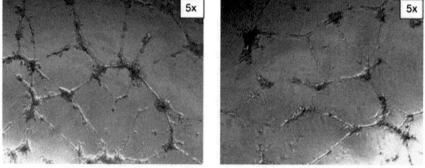

Fig. 13.7 BMP-2 and BMP-4 treatment increases the outgrowth of network structures for human microvascular endothelial cells on Matrigel. (Rothhammer et al., 2007. Reprinted with permission, (C) 2007 Springer Nature)

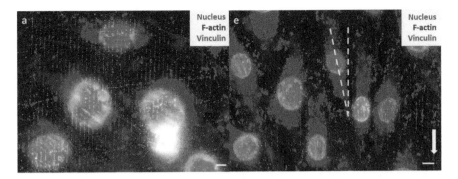

Fig. 13.8 HUVECs are anchored to the nanoparticle–nanowell composite surface through firm focal adhesion (left) and exposed to the shear flow represented in the white arrow (right). Most HUVECs are aligned in the direction of the flow. (Tran et al., 2013. Reprinted with permission, (C) 2013 John Wiley and Sons)

13.7   Alignment of Vascular Endothelial Cells to Flow

Vascular endothelial cells are aligned to the direction of the blood flow, especially so with increased pressure and increased vessel diameter. Figure 13.8 shows the example where HUVECs are patterned on the nanoparticle–nanowell composite surface to ensure firm focal adhesion under strong flow conditions. This schematic has already been discussed in Chap. 8, Sect. 8.4 (and Fig. 8.9). Upon exposing these HUVECs on the nanoparticle–nanowell composite surface to the shear flow, HUVECs are aligned to the direction of flow.

13.8   Laboratory Task 1: Vessel Sprouting by VEGF on Paper-Based Model

In this task, we will demonstrate the vessel sprouting portion of angiogenesis on an OOC. This protocol is a simplified version of Kaarj et al., 2020, *Journal of Biological Engineering* 14:20 (Fig. 13.9). A nitrocellulose chromatography paper is coated with collagen (refer to Chap. 6, Laboratory Task 1). HUVECs will be cultured in a conventional manner and seeded to both sides of the paper. We will place a plastic block at the center of the collagen-coated paper to create a cell-free central region. A magnet is placed on top of the plastic block, and the paper is placed on a metal board, providing a firm attachment of the plastic block to the paper and preventing cell adhesion on the central area. Eventually, HUVECs will be anchored via focal adhesion to the top and bottom portions of the paper, leaving the central area void of cells. This procedure will mimic the two blood vessels in parallel, an OOC mimicking blood vessels and angiogenesis.

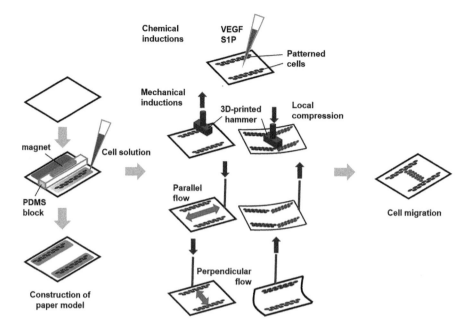

Fig. 13.9 Angiogenesis-on-a-chip is used to simulate the VEGF- and mechanical stimuli-induced angiogenesis. (Kaarj et al., 2020. (C) 2020 Kaarj et al. Open access article distributed under the terms of the Creative Commons Attribution 4.0 International License)

VEGF is then pipetted to the central area, and the paper-based model is placed in a culture medium. After several hours, the paper-based model is fluorescently imaged to monitor vessel sprouting from the existing vessels (top and bottom) to the central area. In Kaarj et al., the paper-based model is also exposed to the repeated movements compression, parallel flow, or perpendicular flow, as shown in Fig. 13.9, using a microcontroller. Kaarj et al. introduced such mechanical stimuli with or without VEGF or other angiogenesis stimulators. This laboratory task will not address such mechanical stimulation, although it will be beneficial to understand the underlying principles.

Objective 1. HUVEC Culture

1. Culture HUVEC in a CO_2 incubator (37 °C and 5% CO_2) using DMEM, supplemented with 10% v/v fetal bovine serum. Also add 0.2% v/v 250 µg/mL amphotericin B (antifungal) and 0.1% v/v 50 mg/mL gentamycin sulfate (antibiotic). Use T-75 or T-25. It is okay to use rat vascular endothelial cells (RVECs). Culture until 90% confluency. Resuspend in a final concentration of 2×10^6 cells/mL.

Objective 2. Cell Patterning on Paper

2. Secure nitrocellulose paper with an average pore size of 10–20 μm. Cut into 11 mm × 15 mm pieces.

3. Add 0.1 mL of 50 μg/mL rat tail collagen type I (= 5 μg collagen) to the entire paper surface and leave it for 1 h. Wash twice with PBS using a wash bottle.

4. Prepare 15 mm × 5 mm × 5 mm plastic block (this can be 3D printed). Place this block at the center of the collagen-coated paper, as shown in Fig. 13.9. Add a magnet on top of the block and place the paper on a metal surface.

5. Add 10 μL of cell solution on each side of the paper. You can pattern just one side of the paper (10 μL) or both sides (10 μL each; 20 μL total). Leave it for 15 min.

6. Remove magnet and plastic block. Place the paper model in a petri dish. Add 3 mL of the media mentioned in step 1.

7. Culture under a static condition (37 °C and 5% CO_2) for 24 h. HUVECs should be covered in monolayer.

Objective 3. Addition of VEGF

8. Take the paper model out from the media. Pipette-add 0.1% v/v 50 ng/mL VEGF to the center. Place the paper model back to the petri dish and add 3 mL of the media. Static culture for another 1 h, and if possible, up to 5 h.

Objective 4. Fluorescence Imaging

9. Image the HUVECs using DAPI and phalloidin-TRITC staining (Fig. 13.10).

13.9   Laboratory Task 2 (Optional): Vessel Sprouting by Mechanical Stimuli on Paper Model

Kaarj et al. demonstrated a microcontroller (Arduino; can be replaced with Raspberry Pi) to provide repeated mechanical stimuli to the paper model. Figure 13.11 shows the photographs of such mechanical stimulations. A 3D printed plastic hammer moves up and down using a servo stepper motor, connected to a motor controller and a microcontroller. This hammer does not physically touch the paper surface. This system provides repeated local compressions to the center of the paper model.

Similarly, a metal wire is connected to the end of the paper model, and its other end is connected to the same servo stepper motor. Lifting the paper up and down creates the relative media flows across the paper surface in a parallel or perpendicular manner (Fig. 13.9). Both local compression and parallel/perpendicular flows were applied for 5 h in a CO_2 incubator.

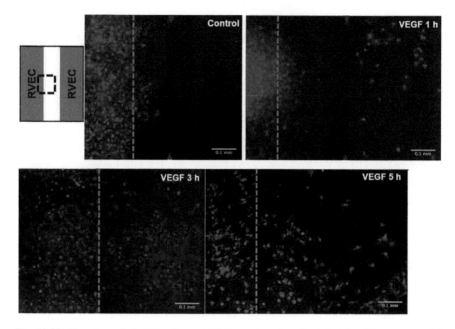

Fig. 13.10 Vascular endothelial cells are initially patterned on the left side of the paper model. After 1 h of pipette-adding VEGF to the central area, vascular endothelial cells migrate toward the center. More migration is observed after 3 h and 5 h, toward vessel sprouting. Blue represents nuclei (via DAPI staining), and red represents actin filaments (via phalloidin-TRITC staining)

Figure 13.12 shows the results of mechanical stimulations. Local compression does induce the migration of vascular endothelial cells, although the resulting sprouting is not continuous. Shear flow shows better results, showing a much-improved extent of migration and continuous sprouting.

Application of both chemical (VEGF) and mechanical stimuli (compression or shear flow) will further improve the extents of sprouting, although not dramatically improved, as demonstrated in Kaarj et al.

Review Questions
1. List the cells in each layer of a typical large-diameter blood vessel, from the lumen to the outside extracellular matrix (vascular endothelial cells - > basement membrane - > vascular smooth muscle cells -> fibroblasts + ECM).
2. What are ephrin and Eph receptors? What are their roles?
3. Two essential growth factors in angiogenesis are VEGF (vascular endothelial growth factor) and FGF (fibroblast growth factor). What are their primary functions?
4. Explain the roles of platelet-derived growth factor (PDGF) and transforming growth factor β (TGF-β) during angiogenesis.
5. What is neuropilin-1, and what is its function toward angiogenesis?
6. List all factors involved in blood coagulation and wound healing that can function as angiogenesis stimulators or inhibitors.

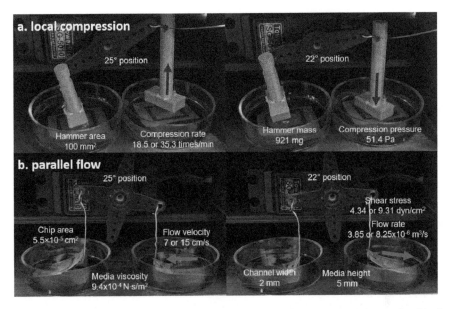

Fig. 13.11 Repeated mechanical stimulations to the paper model toward angiogenesis. (Kaarj et al., 2020. (C) 2020 Kaarj et al. Open access article distributed under the terms of the Creative Commons Attribution 4.0 International License)

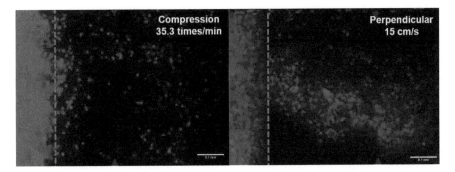

Fig. 13.12 Vessel sprouting by repeated local compression (left) or repeated shear flow (right) on a paper model

7. How can the angiogenesis receptor function as an angiogenesis stimulator or an angiogenesis inhibitor?
8. Compare Ang-1, Ang-2, Tie-1, and Tie-2.
9. Explain how TSP-1, TSP-2, and PAI-1 function as angiogenesis inhibitors.
10. Why is VEGF a popular choice for inducing vascularization to TE transplants?
11. How do anti-VEGFs (including bevacizumab) work as angiogenesis inhibitors? How do they work to treat cancer?

12. What is HUVEC (human umbilical vein endothelial cell), and why is it popular in angiogenesis and TE studies?
13. What is BMP-4 and its role in angiogenesis?

References

Adams, R. H., & Klein, R. (2000). Eph receptors and ephrin ligands: Essential mediators of vascular development. *Trends in Cardiovascular Medicine, 10*, 183–188. https://doi.org/10.1016/S1050-1738(00)00046-3

Bryan, B. A., & D'Amore. (2007). What tangled webs they weave: Rho-GTPase control of angiogenesis. *Cellular and Molecular Life Sciences, 64*, 2053–2065. https://doi.org/10.1007/s00018-007-7008-z

Kaarj, K., Madias, M., Akarapipad, P., Cho, S., & Yoon, J.-Y. (2020). Paper-based in vitro tissue chip for delivering programmed mechanical stimuli of local compression and shear flow. *Journal of Biological Engineering, 14*, 20. https://doi.org/10.1186/s13036-020-00242-5

Pill, K., Hofmann, S., Redl, H., & Holnthoner, W. (2015). Vascularization mediated by mesenchymal stem cells from bone marrow and adipose tissue: A comparison. *Cell Regeneration, 4*, 8. https://doi.org/10.1186/s13619-015-0025-8

Rajabi, M., & Mousa, S. A. (2017). The role of angiogenesis in cancer treatment. *Biomedicine, 5*, 34. https://doi.org/10.3390/biomedicines5020034

Rothhammer, T., Bataille, F., Spruss, T., Eissner, G., & Bosserhoff, A. K. (2007). Functional implication of BMP4 expression on angiogenesis in malignant melanoma. *Oncogene, 26*, 4158–4170. https://doi.org/10.1038/sj.onc.1210182

Tran, P. L., Gamboa, J. R., McCracken, K. E., Riley, M. R., Slepian, M. J., & Yoon, J.-Y. (2013). Nanowell-trapped charged ligand-bearing nanoparticle surfaces – A novel method of enhancing flow-resistant cell adhesion. *Advanced Healthcare Materials, 2*, 1019–1027. https://doi.org/10.1002/adhm.201200250

Chapter 14
Advanced Topics

We have now arrived at the final chapter. We have covered all fundamental theories and laboratory techniques necessary for tissue engineering applications in Chapters 1 through 10. We have also covered the well-established tissue engineering applications, namely organ-on-a-chip (OOC) and tissue-engineered (TE) skin transplant, in Chaps. 11 and 12. Vascularization is the first "advanced" topic of tissue engineering, previously considered optional but recently becoming very important. In this chapter, we will cover more advanced examples of TE transplants.

While it is possible to design and construct "any" tissues and organs with tissue engineering, some tissues and organs are noticeably easier to build and implement while others are not. We will focus on the "easy" examples of such tissue-engineered tissues and organs: cartilage, bone marrow, cardiac "patch," immunoisolated pancreas, etc.

Inquiry 1. Why do you think the above tissues and organs are easier to be made into TE transplants?

Inquiry 2. What organs in the human body will be challenging to be constructed with tissue engineering? Can you briefly explain why?

14.1 Cartilage Tissue Engineering

Cartilages can be found in many parts of the human body and provide structural support along with bones. Cartilages are classified into three types:

- *Hyaline cartilage*, found in joints, ribs, nose, trachea, and larynx
- *Elastic cartilage*, found in ear, epiglottis, and larynx
- *Fibrous cartilage*, found in intervertebral disc

Cartilages can be "worn out" over time, especially by aging or excessive use. Such damages typically occur at the joints. Figure 14.1 shows a typical *joint*, where

© Springer Nature Switzerland AG 2022   249
J.-Y. Yoon, *Tissue Engineering*, https://doi.org/10.1007/978-3-030-83696-2_14

Fig. 14.1 Hyaline
cartilages found at the
joint. Note that the two
bones are connected
directly by *ligaments* or
through muscles by
tendons. The hyaline
cartilages are exposed to
the *synovial cavity*, filled
with *synovial fluid*

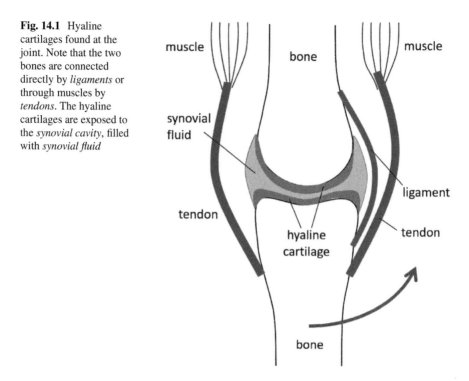

two long bones *articulate* with each other. The heads of two long bones are covered
with cartilages. These cartilages, found at the joint, are hyaline cartilages. They are
also called *articular cartilages*, as the two long bones move over each other, that is,
articulate.

Cartilage consists primarily of extracellular matrix (ECM) and *chondrocytes* are
dispersed at a low density. It has no blood supply, nerves, or lymphatics. It must be
able to resist compression dynamically. Like other tissues, cartilage's ECM is made
from collagens and GAGs. Collagens are significant backbones of cartilage and
provide structural support. GAGs in the cartilage are classified into *nonaggregating
GAGs* and *aggregating GAGs*. Nonaggregating GAGs in the cartilage consist of
keratin sulfate (*KS*), *chondroitin sulfate* (*CS*), and *dermatan sulfate* (*DS*).
Aggregating GAGs are made from binding nonaggregating GAG to *hyaluronic acid*
(*HA*). Hyaluronic acid is critical in cartilage, and chondrocytes have a cell–surface
receptor, *CD44*, that can specifically bind to hyaluronic acid.

Chondrocyte's primary function is the production of ECM. It also has highly
developed actin filaments. Chondrocyte's proliferation is limited; it can proliferate
(expand) only up to 10–30 times in vitro. As such, cartilage's self-repair capacity is
limited. Some chondrocytes proliferate better than others and exhibit stem cell
properties. However, it is challenging to locate such cells. Therefore, when the cartilage is damaged, it does not recover well; hence a tissue-engineered transplant
is needed.

In the past, surgeons drilled a hole from the damaged articular cartilage down to the bone. It will attract bone marrow cells [*mesenchymal stem cells (MSCs)*], and they migrate and differentiate into new chondrocytes, creating new cartilage. Unfortunately, the resulting cartilage is fibrous cartilage that is less dense and less strong than hyaline cartilage.

The tissue engineering approach is a better solution. Through tissue biopsy, healthy chondrocytes are collected from the undamaged articular cartilage. Since the patient's cells are used, this process is considered *autologous*. Through enzymatic digestion, chondrocytes are isolated. These cells are cultured in vitro. Luckily, chondrocytes grow pretty well in vitro, although their doubling time is still quite long. After the desired cell density is achieved, cells are detached from the surface (e.g., using trypsin-EDTA). The isolated chondrocytes are then injected into the damaged area. This process is called *autologous chondrocyte transplantation* (*ACT*), also known as *autologous chondrocyte implantation* (*ACI*), graphically illustrated in Fig. 14.2. While not shown in the figure, a *periosteal flap* is sutured over the damaged area to provide additional protection.

While ACT (ACI) has worked, it has many limitations:

(1) Chondrocytes have a long doubling time.
(2) Chondrocytes can expand (proliferate) only up to 10–30 times (in most cases, 10 times).
(3) Lack of scaffold leads to insufficient strength and compression resistance; thus, ACT (ACI) works only for small lesions.

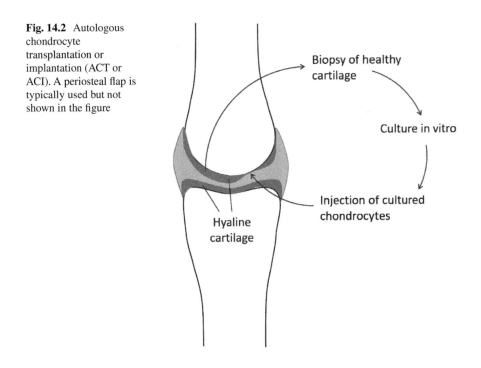

Fig. 14.2 Autologous chondrocyte transplantation or implantation (ACT or ACI). A periosteal flap is typically used but not shown in the figure

Biopsy of healthy cartilage

Culture in vitro

Injection of cultured chondrocytes

Hyaline cartilage

To address (1) and (2), stem cells have been studied instead of autologous chondro-cytes. Embryonic stem cells (ESCs), induced pluripotent stem cells (iPSCs), and mesenchymal stem cells (MSCs) have been investigated for cartilage tissue engi-neering. *Bone morphogenetic protein 2* (*BMP-2*) and *BMP-4* have been identified to induce differentiation into chondrocytes. As always, correct differentiation has been a significant challenge. To address (3), gels made from collagen, fibrin, and decalci-fied bone have been attempted. Collagens are the primary backbones of the chon-drocyte's ECM and provide suitable material characteristics. Fibrins are blood clots also work great in terms of new tissue development. Decalcified bones (bone's ECM are made from collagens and calcium minerals) provide even better material properties. Gels made from biodegradable polymers have also been tested. These polymers include polylactic acid (PLA), polyglycolic acid (PGA), and their copoly-mer PLGA (Sect. 6.9 and Fig. 6.10). Various synthetic polymers can also be used to construct TE chondrocyte scaffolds, and polyesters are preferred considering their strong compression resistance. Finally, gels made from hyaluronic acid receive sig-nificant attention in recent years due to their compatibility with chondrocytes (via CD44) and compositional similarity to hyaline cartilages.

14.2 Bone Marrow Transplantation

While it is possible to construct a TE bone transplant (and they are being investi-gated by many), bones are typically replaced by metallic materials. Metallic bone's clinical outcome is much better than other implants. However, there is a different need for tissue engineering – bone marrow.

Bone marrows can be found inside the cells. Figure 14.3 graphically illustrates the internal structure of a long bone. The ends of a long bone are *spongy bones* (also called *cancellous bones*), whose form is spongy-like. Most of a long bone is *com-pact bone* (also called *cortical bone*). *Red marrows* can be found within the pores of a spongy bone. *Yellow marrows* can be found within the core of a compact bone. Bone marrows are filled with stem cells (primarily hematopoietic stem cells that can differentiate into various blood cells) and mesenchymal stem cells (MSCs). Bone marrow is the body's most prolific organ, generating approximately 400 billion cells per day. It can regenerate in about 2–3 days.

Several diseases can cause reduced stem cell production in the bone marrow, for example, systemic infection, cancer, and *sickle cell anemia*. In addition, chemo-therapy and radiation therapy used to treat cancer can also cause reduced bone mar-row production. *Bone marrow transplantation* was developed and has successfully been used to overcome this problem. Bone marrow is harvested from a patient before chemotherapy (or radiation therapy). These cells are reimplanted sometime after the chemotherapy (or radiation therapy). The harvested cells can also be pro-liferated in vitro, and it thus becomes a tissue engineering method.

Fig. 14.3 Bone marrows
in a long bone

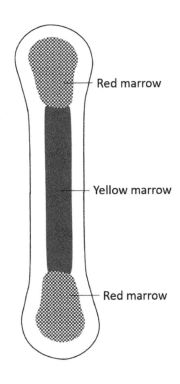

Red marrow

Yellow marrow

Red marrow

14.3 Cardiac Patches

In Chap. 1, we have briefly learned the use of decellularized matrix toward recon-
structing a whole heart. While it is ambitious, there are multiple issues with this
method, as explained in the earlier chapters. The use of a well-designed scaffold
would be the ultimate solution. However, the heart is quite a complicated organ that
is very difficult to be recapitulated with tissue engineering.

One tissue engineering approach has popularly been investigated: cardiac patches
to treat *myocardial infarction* (*MI*). MI, commonly known as *heart attack*, is the
most common form of heart disease (Fig. 14.4). When a plague is built up inside the
coronary artery, blood supply to the *myocardium* (heart muscle tissue) is suspended,
causing damage to the myocardium. A balloon can be surgically inserted and
inflated to expand the occluded coronary artery, called *balloon angioplasty* or simply *angioplasty*. More recently, a cylindrical metallic mesh (*stent*) is added to angioplasty and permanently left inside the coronary artery. In some cases, damages
made to the myocardium can be very severe, requiring a tissue engineering solution.
A cardiac patch can be the ultimate solution for this case built with tissue engineering technology.

Cardiac patches are typically made with tissue engineering technology to treat
the scars in the myocardium after MI. Figure 14.5 shows one such example.

Fig. 14.4 Myocardial infarction (MI) or heart attack. Illustration by Blausen Medical Communications, Inc. and placed in the public domain. (Accessed May 2021 from https:// commons.wikimedia.org/ wiki/File:Blausen_0463_ HeartAttack.png)

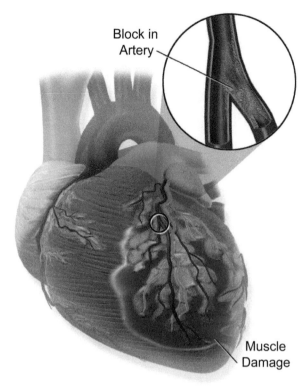

Heart Attack

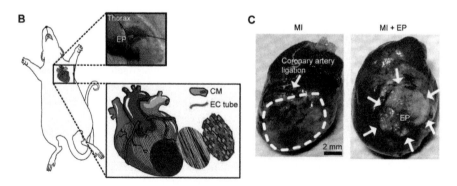

Fig. 14.5 An example of a cardiac patch. Electrospun fiber patch (EP) was used as a TE scaffold. (Lin et al., 2014. Reprinted with permission, (C) 2014 Royal Society of Chemistry)

Cardiomyocytes (primary cells in the myocardium) are harvested and cultured in vitro. A microporous polymer scaffold is designed and constructed. Cultured cardiomyocytes are then seeded and further grown on this scaffold. This TE cardiac patch is then surgically stitched to the scar in the myocardium. As it is just a patch, pre-vascularization may be unnecessary. Post-transplantation vascularization from the nearby tissue can be expected. Various materials can be used as scaffolds for cardiac patches, including decellularized matrix, collagen, gelatin, fibrin, chitosan, hyaluronic acid, PCL, PLA, PGA, PLGA, etc.

Issues of cardiac patches include:

– Cardiac patches may not adequately respond to the electrical stimulation (to induce heartbeat) as the healthy myocardium does.
– Pre-vascularization may still be necessary if the patch size is big.

14.4 Immunoisolated Pancreas

In Chap. 1, we learned the TE pancreas transplant as a TE organ transplant example. Pancreas's primary function is the release of protein *insulin*, regulating the blood glucose level. Problem in insulin causes a disease called *diabetes*. While diabetes can be controlled with drugs, diet, and exercise, self-injection of insulin is often necessary to control diabetes. In severe diabetes cases, this practice may be insufficient, and the TE pancreas transplant becomes essential. *Pancreatic cancer* is another disease that can significantly benefit from the TE pancreas transplant as its mortality is one of the highest among all cancers.

The primary cell in the pancreas is the *β-islet cells*, responsible for insulin production. In severe diabetes and pancreatic cancer, healthy β-islet cells are difficult to locate from the patient (i.e., autologous cells). The best practice is to harvest them from a healthy donor, proliferate them in a laboratory (i.e., in vitro), and transplant the proliferated cells to the patient. In this manner, minimum harm is done to the donor. While a well-matched donor should be identified to minimize immune rejection, such β-islet cells from a donor will eventually be recognized by the patient's immune system. Therefore, the patient should be administered the immunosuppressant drug indefinitely. A TE scaffold may resolve this issue, which is made by a semipermeable membrane. Figure 14.6 shows several examples of such semipermeable membranes: (A) a cylindrical tube (a macrocapsule), (B) a push-pull device with refillable oxygen in it, and (C) a sphere (a microcapsule). The donor β-islet cells inside this scaffold are isolated from the patient's immune response. At the same time, the semipermeable membrane allows the movements of oxygen, nutrients, wastes, and protein products, including insulin. This procedure is called *immunoisolation*, which has been applied to other tissue engineering applications.

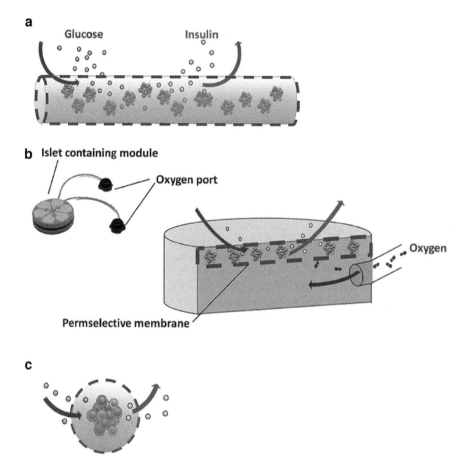

Fig. 14.6 (Identical to Fig. 1.6) Immunoisolation methods can be used toward TE pancreas transplant. Polymeric approaches to reduce tissue responses against devices applied for islet-cell encapsulation. (Hu & de Vos, 2019. (C) Open access article distributed under the terms of the Creative Commons Attribution 4.0 International License)

14.5   Kidney Tissue Engineering

A *kidney* is responsible for the filtration of blood and the excretion of wastes into the urine. It is made from many *nephrons* (Fig. 14.7). A kidney is one of the early demonstrations in organ-on-a-chip (OOC), as explained in Chap. 11. While a single channel may be enough in mimicking the kidney function, a bundle of channels is necessary to recapitulate the whole operation of the kidney to be used as a transplant.

When there is a severe problem in the kidney (called *kidney failure* or *renal failure*), one may consider an organ transplant from a healthy donor or a deceased

Fig. 14.7 (Identical to Fig. 11.5) Nephron, a unit structure of a kidney. The image was made by madhero88 in March 2010 and placed in the public domain. (Accessed March 2021 from https://commons.wikimedia.org/wiki/File:Physiology_of_Nephron.png)

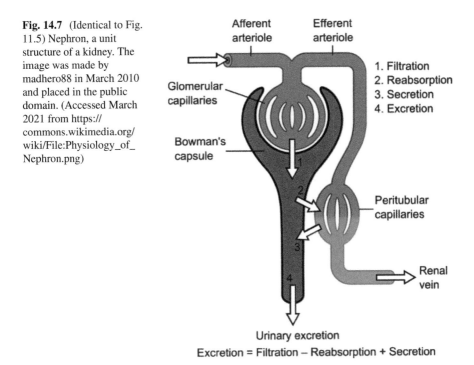

donor. Humans have two kidneys and can survive with a single kidney. However, the patient's immune system will recognize this transplant as foreign, and the patient should be administered an immunosuppressant drug indefinitely. Before such organ transplantation, the patient must receive *hemodialysis* regularly (typically several times a week, each lasting several hours) in a hospital. Patient's blood is pumped out to a hemodialysis machine, filtered, and returned to the patient. The filtration module is a bundle of *hollow fibers*, mimicking the structure of a kidney. We have already learned the use of this hollow fiber module as a tissue engineering bioreactor (Fig. 14.8).

The hemodialysis machine is an ex vivo device and cannot be considered a TE device as it lacks the cells and their proliferation. However, suppose the hollow fiber module is seeded with kidney epithelial cells. In that case, it may be considered a TE kidney device (still ex vivo), which may function better than the existing hemodialysis machine. In addition, such a TE kidney device may be transplanted to the patient, becoming an actual TE transplant.

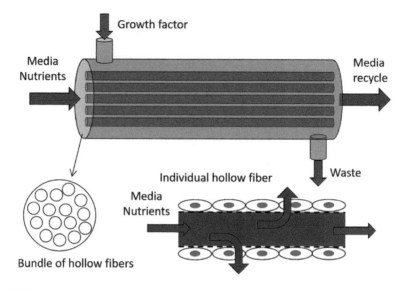

Fig. 14.8 (Identical to Fig. 10.8) Hollow fiber bioreactor

14.6   Other TE Transplants

While not covered in this chapter, it is also possible to develop other TE organ transplants, including bladder, blood vessel, heart, liver, lung, nerve, small intestine, etc.

Review Questions
1. Identify three types of cartilages and where they can be found in the human body.
2. What is articular cartilage, and where can you find it? To which type of cartilage does the articular cartilage belong?
3. What is a chondrocyte and what is its function?
4. How do chondrocytes bind to GAGs?
5. Describe the ACI process. Discuss its current limitations.
6. What are the benefits of stem cells in cartilage tissue engineering? What are the current limitations of using stem cells in cartilage tissue engineering?
7. When do you need a TE scaffold in cartilage tissue engineering?
8. Describe the bone marrow transplantation process. When is it needed?
9. Describe MI and how angioplasty and stent can resolve MI.
10. What is a cardiac patch and when is it needed?
11. What is a β-islet cell and what is its function?
12. What is immunoisolation and why is it needed?
13. Describe the hemodialysis procedure.
14. How does the hollow fiber module filter blood?
15. How can you convert the hollow fiber dialysis module into a TE device?

References

Lin, Y. D., Ko, M. C., Wu, S. T., Li, S. F., Hu, J. F., Lai, Y. J., Harn, H. I. C., Laio, I. C., Yeh, M. L., Yeh, H. I., Tang, M. J., Chang, K. C., Su, F. C., Wei, E. I. H., Lee, S. T., Chen, J. H., Hoffman, A. S., Wue, W. T., & Hsieh, P. C. H. (2014). A nanopatterned cell-seeded cardiac patch prevents electro-uncoupling and improves the therapeutic efficacy of cardiac repair. *Biomaterials Science, 2,* 567–580. https://doi.org/10.1039/C3BM60289C

Hu, S., & de Vos, P. (2019). Polymeric approaches to reduce tissue responses against devices applied for islet-cell encapsulation. *Frontiers in Bioengineering and Biotechnology., 7,* 134. https://doi.org/10.3389/fbioe.2019.00134

Index